12th International Conference on Asia Agriculture and Animal (ICAAA 2025)

BIO Web of Conferences Volume 164 (2025)

Osaka, Japan
17-19 January 2025

ISBN: 979-8-3313-1550-4

Printed from e-media with permission by:

Curran Associates, Inc.
57 Morehouse Lane
Red Hook, NY 12571

Some format issues inherent in the e-media version may also appear in this print version.

For additional information, please contact EDP Sciences – Web of Conferences
at the address below.

EDP Sciences – Web of Conferences
17, Avenue du Hoggar
Parc d'Activité de Courtabœuf
BP 112
F-91944 Les Ulis Cedex A
France

Phone: +33 (0) 1 69 18 75 75
Fax: +33 (0) 1 69 28 84 91

contact-edps@webofconferences.org

Additional copies of this publication are available from:

Curran Associates, Inc.
57 Morehouse Lane
Red Hook, NY 12571 USA
Phone: 845-758-0400
Fax: 845-758-2633
Email: curran@proceedings.com
Web: www.proceedings.com

TABLE OF CONTENTS

FOOD ANALYSIS AND NUTRITIONAL HEALTH

ICAAA 2025

2025 12th International Conference on
Asia Agriculture and Animal

January 17-19, 2025 Osaka, Japan

https://doi.org/10.1051/bioconf/202516400001

Preface

2025 12th International Conference on Asia Agriculture and Animal (ICAAA 2025) was successfully held during January 17-19, 2025 in Osaka, Japan. As we gather once again for this prestigious event, we are excited to explore and discuss the latest advancements, innovations, and challenges in agriculture and animal development of Asia. ICAAA is an annual event initiated since 2016 and will be held every year. We believe that the participants have gained rich knowledge and experience through this conference, which will positively impact on their work and learning in their respective fields.

The papers included in these proceedings are a compilation of the accepted submissions and represent the key outcome of the conference. Each submission underwent rigorous peer review to ensure that only the most outstanding and innovative research results are included in these proceedings.

Finally, we would also like to express our gratitude to the organizing committee, session chairs, reviewers, and everyone involved in making this conference a resounding success. Your dedication and hard work have been instrumental in creating an inclusive and enriching environment for the exchange of ideas and the cultivation of new connections.

I hope that these proceedings will be beneficial as a valuable reference and that the information in this volume will contribute to further advancements in both research and industry in the field of agriculture and animal science.

Kokyo Oh,
Conference Chair
Center for Environmental Science in Saitama, Japan

Committee

Conference Chairs

Kokyo Oh, Center for Environmental Science in Saitama, Japan
Akira Kondo, Osaka University, Japan

Program Chairs

Satoshi Soda, Ritsumeikan University, Japan
Hyunook Kim/ University of Seoul, Korea
Win Chaeychomsri, Kasetsart University, Thailand
Arifin Noor Sugiharto, University of Brawijaya, Indonesia

International Technical Committee

Tumen Wuliji, Lincoln University, Missouri, USA
Turgut AYGÜN, Bingöl University, Turkey
Saowapar Khianngam, Silpakorn University, Thailand
Attapol Tiantong, Silpakorn University, Thailand
Jet Saartje Mandey, Sam Ratulangi University, Indonesia
Meera Srivastava, Govt. Dungar College, India
Abubakar Salihu, Modibbo Adama University of Technology Yola, Nigeria
Saiful Irwan Zubairi, Universiti Kebangsaan Malaysia (UKM), Malaysia
Anoop Kumar Srivastava, Indian Council of Agricultural Research- Central Citrus Research Institute, India
Hsiu-Lin Huang Chung Hsing University, Taiwan
Mohammed Jabar ABDULRADH, The University of Queensland, Australia
HALIMATUN YAAKUB, Universiti Putra Malaysia, Malaysia
Shahram Nanekarani, College of Agriculture & Natural Resources, IAU, Branch Broujerd, Iran
Juni Sumarmono, Universitas Jenderal Soedirman, Purwokerto, Indonesia
Fatiha HAKIMI, Hassan II Institute of Agronomy and Veterinary Medicine, Morocco
Suraya Mohammad Salleh, Universiti Putra Malaysia, Malaysia
Amjed Ali, University of Sargodha, Pakistan
Loh Khye Er, Tunku Abdul Rahman University of Management and Technology, Malaysia
Frisco Nobilly, Universiti Putra Malaysia,Malaysia
Mohamad Zaihan Bin Zailan, Universiti Malaysia Sabah, Malaysia
Sharmila Ahmad, Universiti Putra Malaysia, Malaysia

Statement of Peer review

In submitting conference proceedings to *Web of Conferences*, the editors of the proceedings certify to the Publisher that

1. They adhere to its **Policy on Publishing Integrity** in order to safeguard good scientific practice in publishing.
2. All articles have been subjected to peer review administered by the proceedings editors.
3. Reviews have been conducted by expert referees, who have been requested to provide unbiased and constructive comments aimed, whenever possible, at improving the work.
4. Proceedings editors have taken all reasonable steps to ensure the quality of the materials they publish and their decision to accept or reject a paper for publication has been based only on the merits of the work and the relevance to the journal.

Title, date and place of the conference

2025 12th International Conference on Asia Agriculture and Animal (ICAAA 2025)
January 17-19, 2025 in Osaka, Japan

Proceedings editor(s):

Kokyo Oh, Center for Environmental Science in Saitama, Japan

Date and editor's signature 23 Feb., 2025

Comparative effects of butyric acid blend with valeric acid and tributyrin on growth performances and carcass quality of broiler chicken

Theerawit *Poeikhampha*[*1], Sombat *Prasongsook*[1], Savek *Kiatsomphob*[1], Panatda *Bungsrisawat*[1] and Warod *Buranasiri*[1]

[1]Department of Animal Science, Faculty of Agriculture, Kasetsart University, Bangkok 10900, Thailannd.

Abstract. The use of alternative feed additives as a replacement for antibiotic growth promoters has been focused on increasing the growth efficiency of broiler. This study was conducted to compare the effects of butyric acid blend with valeric acid and tributyrin on growth and carcass in broiler. A total of 630 male broiler chickens (ROSS 308) was divided into 3 groups and distributed into 6 pens with each pen consisting of 35 birds following a Completely Randomized Design. Throughout the 35 days of the feeding trial, all birds received diets: 1) Control diet, 2) +Butyric acid blend with Valeric acid and 3) +Tributyrin. The results indicated that birds received butyric blend with valeric acid or tributyrin showed significantly improved feed conversion ratio ($p<0.05$). Birds receiving butyric blend with valeric acid tend to increase daily weight gain. Feed intake seems to be decreased by tributyrin. Conversely, the birds received butyric blend with valeric acid or tributyrin supplementation not influenced carcass yield and foot-pad score. Therefore, the supplementation of butyric blend with valeric acid or tributyrin results in improved growth performance without the negative effect on carcass quality in broiler chicken.

[*] Corresponding author: agrtrw@ku.ac.th

1.Introduction

Recently, the use of feed additives in stimulating the growth and improving carcass quality of chickens has been increasingly studied because of the rapid development of broiler breeds and broiler farming management. Since 2006, the European Union had banned antimicrobial growth promoter (AGP) because antimicrobial resistance (AMR) problems in human food safety. The withdrawing of AGP from routine production and seeking for AGP replacer have become a worldwide target for sustainable industry development [2] Withdrawal of AGP significantly increases the incidence of diseases and reduces animal growth performance and efficiency. Due to such consequences, it is very important in finding the optimal alternatives for AGP.

Butyric acid is known to be involved in the mucosal immune response, and it has an anti-inflammatory effect in animals. Butyrate can increase the rate of tight junction which suggests butyrate helps to maintain gut integrity [9] Another study observed increasing of machine production, a secreted component of intestinal mucosal barriers, in human intestinal epithelial cells incubated with SCFA, including butyrate [13] Butyrate supplementation for broiler chickens is known to have beneficial properties such as reducing Salmonella colonization and increasing antimicrobial peptide transcription [3,12] Moreover, in healthy broilers, butyrate supplementation leads to reduced weight gain that is caused by a lower nutrient diet [1]

Valeric acid is a straight-chain saturated fatty acid containing five carbon atoms. It is a short-chain fatty acid and a straight-chain saturated fatty acid. The valeric might give the positive effects when supplementation in the diet, however, Effects of valeric acid on intestinal health have been poorly investigated [7] This study was conducted to compare the effect of butyric acid blended valeric acid and tributyrin on the growth performance and the carcass quality of broiler chicken. To find out a suitable guideline for selecting animal feed additives which focus on improving growth performance and carcass yield for broiler industry.

2.Materials and Methods

This study was conducted at the Animal Research Farm, Department of Animal Science, Faculty of Agriculture, Kasetsart University, Thailand. The experimental animals were kept, maintained, treated and handled in adherence and accordance with accepted standards for the humane treatment of animals. The number of Approval for Animal Care and Use of Scientific Research, Kasetsart University is ACKU65-AGR-016.

2.1 Animal and management

Six hundred- and thirty-day-old male chicks (ROSS 308 strain) were obtained from a commercial hatchery. The birds were randomly assigned according to their initial body weights to one control group and 3 treatment groups. Each group was replicated into 6 pens with each pen consisting of 35 birds. The birds were given the same management and vaccination program as those done in commercial farms. Water and feed were provided *ad libitum* throughout the experiment. The water supply in each pen was equipped with a water bell from 1-10 days and nipple from 11-42 days. The birds in a control group received the basal diet and water (without any supplementation) (T1). Basal diet supplemented with butyric and valeric acid dosages 500 g/ton in starter, 500 g/ton in grower and 250 g/ton in finisher (T2) and Basal diet supplemented with tributyrins dosages 500g/ton in starter, 250 g/ton in grower and 250 g/ton in finisher (T3).

Table 1. Feed ingredients and nutrients composition of the basal diet (control diet).

Item	Starter (1-10 days)	Grower (11-24 days)	Finisher (25-35 days)
Corn	55.10	41.62	40.99
Wheat	-	10.00	15.00
Rice bran defatted	-	5.00	5.00
Corn DDGs	-	8.00	8.00
Palm oil	2.78	4.97	5.99
SBM 48%	37.63	21.13	15.94
Rape seed Meal	-	4.00	4.00
Feather meal	-	2.00	2.00
L-Lysine HCL	0.19	0.35	0.35
L-Threonine	0.09	0.08	0.06
Salt	0.35	0.35	0.35
Calcium carbonate	1.21	1.02	1.08
MDCP 21%	1.80	0.73	0.53
Vitamin and Mineral Premix+	0.24	0.24	0.24
Choline chloride 60%	0.28	0.27	0.25
Phytase and Xylanase	-	0.01	0.01
Total	100.00	100.00	100.00
Nutrient (Calculation)	Starter (1-10 days)	Grower (11-24 days)	Finisher (25-35 days)
Metabolize energy for Poultry (kcal/kg)	3,000.00	3,100.00	3,200.00
Protein (%)	23.00	21.50	19.50
Fat (%)	5.27	7.81	8.82
Fiber (%)	2.56	3.58	3.47
Calcium (%)	0.96	0.81	0.78
Total Phosphorus (%)	0.72	0.58	0.52
Avail. P for Poultry (%)	0.48	0.44	0.39

Salt (%)	0.34	0.34	0.34
Animal acids (Digestible, %)			
Arginine	1.44	1.22	1.08
Histidine	0.58	0.50	0.45
Isoleucine	0.90	0.79	0.70
Leucine	1.81	1.67	1.53
Lysine	1.28	1.15	1.02
Methionine	0.63	0.53	0.48
Methionine +Cysteine	0.95	0.87	0.80
Phenylalanine	1.05	0.95	0.86
Phenylalanine +Tyrptophan	1.77	1.36	1.20
Threonine	0.86	0.77	0.68
Tryptophan	0.25	0.21	0.19
Valine	0.98	0.92	0.83
Choline (mg/kg)	1,700.00	1,600.00	1,500.00

+The vitamin and mineral premix contained in the diet: vitamin A, 8,820 kIU/kg; vitamin D, 3,528 kICU/kg; vitamin E, 13,230 IU/kg; menadione, 1,166 mg/kg; B12, 11 mg/kg; folic acid, 441 mg/kg; choline, 835 mg/kg; d-pantothenic acid, 14,112 mg/kg; riboflavin, 4,410 mg/ kg; niacin, 70,560 mg/kg; thiamine, 2.205 mg/kg; d-biotin, 66 mg/kg; pyridoxine, 1,600 mg/kg; manganese, 68 g/kg; zinc, 68 g/kg; iron, 48 g/kg; copper, 6 g/kg; iodine, 0.6 g/kg; selenium, 0.24 g/kg.

All birds were fed a starter diet until 10 days of age, followed by grower and finisher diet from 11-24 days and 25-42 days of age, respectively. The feed ingredients used, and nutritional composition of experimental diets are shown in Table 1. The diets were formulated to meet the recommended nutritional contained 23% CP and 3,000 ME kcal kg-1 during the starter period, 21.50% CP and 3,100 ME kcal kg-1 during the growing period and 19.50% CP and 3,200 ME kcal kg-1 during the finisher period. The experiment was conducted during the month of June to October 2022 in Bangkok, Thailand. Throughout the duration of the experiment, house temperature and humidity readings were taken and recorded once a day at 13.00.

2.2 Growth performance

Body weight, feed intake and flock mortality in each pen were recorded during all the feeding stages. Mean body weight, weight gain and feed intake were computed to determine feed conversion ratio (FCR).

2.3 Carcass quality measurements

When the experiment ended in the 42 days, 12 birds were randomly chosen from each group and were asphyxiated using CO_2 under atmosphere of less than 2% oxygen (air displaced by CO_2). The hot carcasses were manually eviscerated and weighted thereafter. Abdominal fats were manually removed and weighted. The weight of carcass weight, abdominal fat, wing, inner breast, outer breast, drumstick and thighs were recorded. Footpad lesions score were assigned to one of three classes: score 0 = no lesions, no or very small superficial lesions, slight discolor ration on a limited area of the footpad, mild hyperkeratosis (thickening of the outer layer of the skin) or healed; score 1 = mild lesion, discoloration of the footpad, superficial lesions, dark papillae and hyperkeratosis; score2 = severe lesions, epidermis is affected, ulcers and scabs, sign of hemorrhages or swollen footpads according to [8]

Statistical analysis: All the data was statistically analyzed using ANOVA. The differences between the means of each group were separated by Duncan's Multiple Range Test. Statements of statistical significance are based on p<0.05. All statistical analyses were computed in accordance with the Steel and Torrie (1980) in accordance with the following model:

$$Y_{ij} = \mu + A_i + \varepsilon_{ij}$$

Where;

Y_{ij} is the observation from the i treatment in the j replication (where i = 1,2,3 and j = 1, 2, 3,.,6)

μ is the overall mean of the experiment

A_i is the effect of the i treatment (where i = 1,2,3)

ε_{ij} is the experimental error ~ NID (0, σ^2).

3. Results and Discussion

3.1 Growth performance

The birds received feed supplemented butyric blend with valeric acid or supplemented with tributyrin significantly decreased average feed conversion ratio 1.37+0.03 and 1.36+0.04 when compared with the control group (p<0.05). Birds received feed supplemented butyric blend with valeric acid tended to increase average daily weight gain 74.53+1.88 g. The birds received feed supplemented with tributyrin were shown average feed intake 3467.16+59 g. which tended to decrease in feed intake when compared to the other groups (Table 2).

Table 2. Effects of Butyric Acid blend with Valeric Acid and Tributyrin on growth performances of broiler chickens.

Items	Control	Butyric blend with Valeric acid	Tributyrin	p-value	SEM
Initial body weight (g)	46.91±0.47	47.11±0.54	46.87±0.25	0.60	0.10
Initial Uniformity (%)	91.63±1.21	90.84±2.82	90.48±0.95	0.56	0.42
Uniformity (%)	88.90±1.43	90.37±0.75	90.47±1.77	0.12	0.35
Average daily weight gain (g)	72.42±0.74	74.53±1.88	73.09±1.95	0.10	0.42
Feed intake (g)	3598.82±110.63	3580.55±100.87	3467.16±59.80	0.06	24.98
Feed conversion ratio	1.42±0.04[a]	1.37±0.03[b]	1.36±0.04[b]	<0.05	0.01
Feed cost per gain (baht/g)	24.08±0.69	23.94±0.75	24.02±0.74	0.94	0.16

Data is Mean±SD, a,b Means with different superscripts in the same row are significantly different (*p*<0.05)

This study affirmed the findings of [4] that the birds received tributyrin significantly improved feed conversion ratios (FCR, *p*<0.05) and shown not statical differences in feed intake. Moreover, this study also affirmed the findings of Sizmaz et al. (2022) that the birds received tributyrin supplementation in the feed can reduce feed conversion ratios (*p*<0.05). As a result of the finding of [5] that supplementation of tributyrin in the feed may lead to increasing nutrition absorption in broiler by improved the intestinal morphology as indicated by higher villus height and the ratios of villus height/crypt depth in broilers.

3.2 Carcass quality

The birds received feed supplemented butyric blend with valeric acid or tributyrin shown not different in carcass weight, abdominal fat, wing weight, inner breast weight, drumstick weight, thighs weight and foot-pad score (Table 3).

Table 3. Effects of butyric acid blend with valeric acid and tributyrin on carcass quality of broiler chickens.

Items	Control	Butyric blend with Valeric acid	Tributyrin	*p*-value	SEM
Carcass weight (g)	2124.92±69.52	2148.33±84.95	2130.17±72.80	0.84	12.37
Abdominal fat (g)	31.92±12.94	32.58±24.34	30.33±17.37	0.96	3.05
Wing (g)	193.17±11.49	193.75±16.07	190.92±11.78	0.86	2.16
Inner breast (g)	104.25±5.93	108.58±14.18	103.17±10.10	0.43	1.76
Outer breast (g)	580.17±40.08	565.25±59.46	556.00±99.74	0.70	11.60
Drumstick (g)	246.58±14.13	247.00±16.45	240.92±17.78	0.59	2.66
Thighs (g)	337.83±52.40	328.83±38.26	340.92±22.67	0.74	6.48
Foot pad score	0.67±0.78	0.67±0.65	1.17±1.27	0.33	0.16

Data is Mean+SD

The result of this study in terms of carcass yield was contradicted with the findings of [6] that birds given tributyrin supplementation in the feed can increase eviscerated carcass rate and reduce abdominal fat yield (p<0.05). Therefore, the supplementation of butyric acid, blended with valeric acid, and tributyrin improved feed utilization, but did not affect carcass yield when compared to those fed the control diet.

4. Conclusion

It can be concluded that butyric acid blended valeric acid and tributyrin can improve feed conversion ratio (FCR) and tended to increase average daily weight gain when fed in the diet and it could be used as feed additives for broiler without the negative to the carcass of broiler.

References

1. Bortoluzzi, C., Pedroso, A. A., Mallo, J. J., Puyalto, M., Kim, W. K., & Applegate, T. J. Sodium butyrate improved performance while modulating the cecal microbiota and regulating the expression of intestinal immune-related genes of broiler chickens. Poultry science, 96(11), 3981-3993. (2017).
2. Cromwell, G.L. Why and how antibiotics are used in swine production. Animal Biotechnology, 13:7–27. (2002)
3. Fernández-Rubio, C., Ordonez, C., Abad-González, J., Garcia-Gallego, A., Honrubia, M.P., Mallo, J.J. and Balana-Fouce, R. Butyric acid-based feed additives help protect broiler chickens from Salmonella Enteritidis infection. Poultry science, 88(5), 943-948 (2009)
4. Hansen, V.L., Kahl, S., Proszkowiec-Weglarz, M., Jiménez, S.C., Vaessen, S.F., Schreier, L.L., Jenkins, M.C., Russell, B. and Miska, K.B. The effects of tributyrin supplementation on weight gain and intestinal gene expression in broiler chickens during Eimeria maxima-induced coccidiosis. Poultry Science, 100(4), 100984 (2021)
5. Hu, Q., Yin, F., Li, B., Guo, Y. and Yin, Y. Dietary Tributyrin Administration Improves Intestinal Morphology and Selected Bacterial and Short-Chain Fatty Acid Profiles in Broilers Under an Isocaloric Feeding Regime. Frontiers in Microbiology, 12 (2021)
6. Hu, Q., Yin, F., Yang, L., Li, B., Lei, G., Wang, C., Yin, Y. and Liu, D. Dietary tributyrin intervention improves the carcass traits, organ indices, and blood biomarker profiles in broilers under the isocaloric diets administration. Poultry Science, 101(10), 102061 (2022)
7. Lonneke O, V.D. Karolien, D. Richard, S. Koen, H. Freddy, V.I. Filip. Valeric acid glyceride esters in feed promote broiler performance and reduce the incidence of necrotic enteritis, Poultry Science, 97(7), 2303-231 (2018)
8. Mihai C.S. Van, I. and Georgeta, C. Influence of the dietary protein level on the incidence of footpad dermatitis in broiler chickens. The Indian Journal of Animal Sciences. 83(10):1084-1089 (2013)
9. Peng, L., Li, Z. R., Green, R. S., Holzman, I. R., & Lin, J. Butyrate enhances the intestinal barrier by facilitating tight junction assembly via activation of AMP-activated protein kinase in Caco-2 cell monolayers. The Journal of nutrition, 139(9), 1619-1625 (2009)
10. Sizmaz, O., Barrett, N.W., Lewis, J., Yakout, H. and Persia, M.E. Effect of various concentration of butyric acid on growth performance, intestinal lesion scores, and body composition of broilers raised on used litter. Journal of Applied Poultry Research, 100296 (2022)
11. Steel, R. G. D., & Torrie, J. H. Principles and procedures of statistics, a biometrical approach 2nd Edition McGraw-Hill Publishing Company Publishing Company Toronto (1980)
12. Sunkara, L. T., Achanta, M., Schreiber, N. B., Bommineni, Y. R., Dai, G., Jiang, W., Lamont, S., Lillehoj, H.S., Beker, A., Teeter, G.R. and Zhang, G.

Butyrate enhances disease resistance of chickens by inducing antimicrobial host defense peptide gene expression. PLoS one, 6(11), e27225 (2011)
13. Willemsen, L. E. M., Koetsier, M. A., Van Deventer, S. J. H., & Van Tol, E. A. F. Short chain fatty acids stimulate epithelial mucin 2 expression through differential effects on prostaglandin E1 and E2 production by intestinal myofibroblasts. Gut, 52(10), 1442-1447 (2013)

Effect of free fatty acids supplementation in diet on metabolizable energy growth performance and carcass quality in broiler chickens

Sombat Prasongsook[1], *Chaowit* Rakangthong[1] and *Theerawit* Poeikhampha[1*]

[1]Department of Animal Science, Faculty of Agriculture, Kasetsart University, Bangkok, Thailand

Abstract. Fats are an excellent source of energy and play an important role in the growth of broilers. Using free fatty acids is one of the interesting alternative choices for fats since there is a major metabolic fuel and a part of triacylglycerols. This study was conducted to evaluate the effects of free fatty acids (FFAs) substitution in diet on metabolizable energy, growth performance and carcass quality of broilers. A total of 900 male broiler chicks were divided into 5 groups with 6 replicates of 30 birds each. The birds received a control diet using palm oil as dietary oil supplementation and other groups were substituted with 25, 50, 75 and 100 % FFAs for 35 days. At the end of the feeding trial, the results indicated that all levels of FFAs substitution did not influenced the body weight, feed intake, FCR as well as carcass quality of broilers when compared to the control group. Moreover, there was no significant difference of the metabolizable energy in broilers fed with FFAs. It can be concluded that FFAs is useful as dietary oil supplementation by substitution of palm oil without negative effects on the metabolizable energy, growth performance and carcass quality of broilers.

1 Introduction

Dietary fats and oils, consisting mainly of triacylgecerols, are relatively large molecules and cannot be absorbed intact in the small intestines. The triacylgecerols must be broken down (hydrolyzed) by lipase and colipase, both of which are enzymes produced by the pancreas. The result of the hydrolysis are two free fatty acids and one 2-monoacylglycerol, which can be absorbed by the small intestines. Fats in the diet are a source of essential fatty acids that are important for the growth of broilers

* Corresponding author: agrtrw@ku.ac.th

and also play a vital role in maintaining healthy skin and feathers, insulating body organs against shock, maintaining body temperature, and promoting healthy cell function. Fats are an excellent source of energy, and their inclusion in the diet is the main method of increasing the energy content of poultry diets [1,2]. Various fat sources are available for poultry, such as animal fats (e.g., lard and tallow) and vegetable oils (e.g., palm oil and soybean oil) but these fats may cost the high price. Therefore, the alternative choice is using the other sources of fat such as free fatty acids.

Free fatty acids (FFAs) is a fatty acid which is part of triacylglycerol without glycerol. Fatty acids play a number of key roles in metabolism – major metabolic fuel (storage and transport of energy). To use free fatty acids, it is essential to know the composition of fatty acids in an oil or fat, to identify their characteristics and to know the stability of them [3,4] because [5,6,7] reported that the metabolizable energy was influenced by the unsaturated: saturated (U:S) ratio. Therefore, the objectives of this study were to evaluate the effects of free fatty acids on metabolizable energy, growth performance and carcass quality of broiler chickens.

2 Materials and methods

2.1 Animals and managements

This study was conducted at Animal Research Farm, Department of Animal Science, Faculty of Agriculture, Kasetsart University, Bangkok, Thailand. The experimental animals were kept, maintained and treated in adherence to accepted standards for the humane treatment of animals.

A total of 900 one-day-old male ROSS 308 broiler chicks (average body weight 46.63 g) were used in this trial. Chicks were randomly divided into five treatments and each treatment consisted of six pens (thirty birds per pen). Feed and water were provided ad libitum, while water was supplied by nipple drinkers. All chicks were inoculated with Newcastle disease vaccine and oral vaccination of the infectious bursal disease vaccine on day 10.

The experimental diets (Table 1 and 2) were satisfying the nutritional requirements described in strain's and based on corn and soybean meal. Five dietary treatments were provided to the broilers for 35 days as follows; basal diet (control), and substitution of palm oil by 25, 50, 75, 100 % FFAs, respectively.

The specifications of FFAs that used in the experiment was liquid substance, brown color, 99% minimum of crude fat, 1% maximum of moisture, 70% minimum of free fatty acids, minimum 80 of iodine value, 30 °C of melting point and fatty acid composition: 18% palmitic acid, 5% stearic acid, 60% oleic acid, 14% linoleic acid and 3% others.

Table 1. Composition of the basal diet

Ingredient	Starter	Grower	Finisher
Thai Corn	55.01	56.39	62.35
Palm oil	3.48	5.06	4.74
Soybean meal (48% CP)	36.07	34.00	28.72
L-Lysine HCL 78%	0.23	0.08	0.08
DL-Methionine	0.33	0.25	0.21
L-Threonine	0.08	0.01	0.01
Monodicalciumphosphate21%	2.19	1.92	1.77
Calcium carbonate	1.51	1.25	1.22
Salt	0.30	0.30	0.30
Premix[a]	0.55	0.53	0.40
Choline Chloride 75%	0.25	0.23	0.22
Total	100.00	100.00	100.00
Calculated chemical composition			
Metabolizable energy (MJ/kg)	3,025	3,150	3,200
Protein (%)	22.10	21.00	19.00
Fat (%)	5.75	7.32	7.17
Calcium (%)	1.05	0.90	0.85
Total Phosphorus (%)	0.85	0.78	0.73
Available phosphorus (%)	0.50	0.45	0.42
Salt (%)	0.32	0.32	0.32
Lysine (%)	1.27	1.10	0.97
Methionine (%)	0.94	0.84	0.76
Methionine+Cystein (%)	0.63	0.54	0.48
Threonine (%)	0.83	0.73	0.65
Tryptophan(%)	0.24	0.23	0.20
Valine (%)	0.95	0.91	0.82

[a]Premix content; Composition per 1 kg: Vitamin A 4.8 MIU, Vitamin D3 1.2 MIU, Vitamin E 6,000 IU, Vitamin K3 0.6 g, Vitamin B1 0.6 g, Vitamin B2 2.2 g, Vitamin B6 0.8 g, Vitamin B12 0.004 g, Nicotinic acid 10 g, Pantothenic acid 4.8 g, Folic acid 0.2 g, Biotin 0.048 g,

Manganese 32 g, Zinc 24 g, Iron 16 g, Copper 3.2 g, Iodine 0.2 g, Cobalt 0.04 g, Selenium 0.04 g

2.2 Growth performance

The initial body weight of each chick was recorded at the beginning of the feeding trial (day 1). The body weight, body weight gain and feed intake were measured on the feed switching time basis, which means evaluating on 10, 24 and 35 days in order to calculate average daily gain (ADG) and feed conversion ratio (FCR) and the mortality of broilers were observed.

2.3 Carcass quality

At the end of the finisher period (35 days), 2 birds per replicate (showing closest to the average body weight of the pen) were selected in order to determine the carcass quality as following; live weight, slaughter weight, carcass weight, inner breast weight, outer breast weight, thigh weight, drumstick weight, wing weight and abdominal fat.

2.4 Metabolizable energy

Four days prior to the end of the finisher period (day 31), 4 birds per unit were moved to the metabolic cage for metabolizable energy examination. Birds were fed with the experimental feed containing 0.5% Cr_2O_3 as the external marker and using celiteTM and AIA (Acids Insoluble Ash) as internal markers. Feed was provided for 80% of recent daily feed consumption and clean collection trays were placed beneath each cage and excreta was collected in labeled plastic bags.

The retention time was determined by the excreta color changing from a normal color to a light green color (color from Cr_2O_3). Then excreta were pooled, homogenized, weighed to 100 g and immediately dried for further analysis (dry matter, gross energy and AIA) in order to calculate metabolizable energy [8].

2.5 Statistical analysis

Analysis of variance was conducted using pen means. All statistical analyzes were performed by ANOVA using SAS statistical program. [9] Differences among treatments were determined with Duncan's multiple range test [10]. Statements of statistical significance were based on $P<0.05$. Then the data were analyzed by regression analysis based on the data that was in response to the increasing levels of FFAs in linear and quadratic equations to predict the levels of FFAs.

3 Results

3.1 Growth performance

The growth performances of broilers are shown in Table 2. All broilers remained in good health throughout the trial. All FFAs substitution levels (25%, 50%, 75% and 100%) did not influence the body weight of starter, grower, finisher period and overall trial (35 days) when compared to the control group and linearly increased the FCR of the overall trial. Substitution of FFAs linearly increased the feed intake and FCR in the starter period. However, the feed intakes of the overall trial (35 days) were not significantly influenced by the FFAs

The results of this study indicated that FFAs did not statistical significance (P>0.05), but they suggest a possible negative effects on broiler performances, which is in agreement with the report of Pesti et al. (2002) [11]. The body weight of broilers fed diets containing FFAs in finisher and overall performance were lower than those of birds fed with a control diet, suggesting that the energy in diets with a higher FFAs content was not utilized as efficiently as the energy in diets with a lower FFAs content [12]. Shannon (1979) [13] and Sklan (1979) [14] reported that the absorption of fatty acids was higher in chicks fed triglycerides than in chicks fed FFAs. Their results suggested a need for glycerides for the efficient solubilization and absorption of FFAs. The cause of lower body weight could be that the diets with higher FFAs levels were associated with some chemical, physical, or physiological characteristic that reduced palatability and feed intake [15].

There were many factors shown to affect the utilization of fats, such as the age of the chicken [16-17], the degree of saturation of fatty acids [18-23], and the chain length of the fatty acid [18], as well as positional effects of fatty acids on the triglyceride molecule [19]. Therefore, in this study, the lower body weight of broilers fed FFAs might be caused by the high levels of FFAs and the high ration of U:S.

Table 2. Effect of free fatty acids supplementation in diet on growth performance of broilers.

Item	Free fatty acids					P-Value	SEM	Contrast, P<	
	0%	25%	50%	75%	100%			Linear	Quadratic
Starter (d 1 to 10)									
BW (g/b)	344.41	345.60	342.64	345.27	347.64	0.83	1.22	0.50	0.50
FI (g/b/d)	34.97	35.01	35.36	36.11	36.68	0.15	0.26	0.01	0.48
BWG (g/b)	297.77	298.96	296.31	298.75	301.02	0.83	1.21	0.49	0.50
ADG (g/b/d)	29.78	29.90	29.63	29.88	30.10	0.82	0.12	0.49	0.50
FCR	1.17	1.17	1.19	1.21	1.22	0.27	0.01	0.03	0.80
Grower (d 11 to 24)									
BW (g/b)	1,375.25	1,350.70	1,352.03	1,358.10	1,347.17	0.47	5.19	0.20	0.47
FI (g/b/d)	99.11	96.34	97.75	97.75	97.80	0.40	0.43	0.69	0.25
BWG (g/b)	1,030.84	1,005.09	1,009.08	1,012.73	999.53	0.25	4.58	0.09	0.51
ADG (g/b/d)	73.63	71.79	72.08	72.34	71.40	0.25	0.33	0.09	0.51
FCR	1.35[b]	1.34[b]	1.36[ab]	1.35[b]	1.37[a]	<0.01	0.00	<0.01	0.26
Finisher (d 25 to 35)									
BW (g/b)	2,432.51	2,382.84	2,409.10	2,364.72	2,363.75	0.40	13.01	0.10	0.81
FI (g/b/d)	165.36	163.33	162.94	161.75	163.25	0.69	0.75	0.30	0.34
BWG (g/b)	1,057.27	1,032.14	1,057.07	1,006.62	1,016.57	0.51	11.14	0.19	0.96
ADG (g/b/d)	96.11	93.83	96.10	91.51	92.42	0.51	1.01	0.19	0.96
FCR	1.72	1.74	1.70	1.77	1.77	0.50	0.01	0.23	0.63
Overall (d 1 to 35)									
BW (g/b)	2,432.51	2,382.84	2,409.10	2,36472	2,363.75	0.40	13.01	0.10	0.81
FI (g)	3,556.23	3,495.46	3,514.36	3,508.79	3,531.69	0.50	11.15	0.66	0.14
BWG (g/b)	2,385.87	2,336.19	2,362.46	2,318.10	2,317.13	0.40	13.01	0.10	0.81
FCR	1.49	1.50	1.49	1.52	1.53	0.21	0.01	0.04	0.47

[a,b] Means within a row with different letters differ significantly (P<0.05

3.2 Carcass quality

Substitution of FFAs linearly reduced slaughter weight, carcass weight and outer breast weight (Table 3) and there were non-significant differences in the other carcass yields (inner breast weight, thigh weight, wing weight, drumstick weight, abdominal fat weight and liver weight). Peebles et al. (1999) [24] reported that birds fed with less saturated fatty acid in the diet improved slaughter yield. Since the carcass quality was affected by the fatty acids profile, therefore, the carcass quality had no significant difference in broilers fed with FFAs.

Table 3 Effect of free fatty acids supplementation in diet on carcass quality of 35 days old broilers.

Item (g)	Free fatty acids					P-Value	SEM	Contrast, P<	
	0%	25%	50%	75%	100%			Linear	Quadratic
Live weight	2,369.17	2,309.25	2,321.67	2,287.00	2,290.75	0.25	12.75	0.05	0.45
Slaughter weight	2,205.08	2,152.50	2,163.92	2,115.92	2,119.83	0.11	12.04	0.02	0.57
Carcass weight	1,935.50	1,890.42	1,901.83	1,855.17	1,848.75	0.09	11.41	0.01	0.83
Inner breast weight	82.67	84.08	84.5	83.42	80.83	0.82	1.01	0.57	0.29
Outer breast weight	420.75	418.33	406.67	395	398.42	0.20	4.21	0.03	0.73
Thigh weight	290.83	279.50	291.17	279.67	277.00	0.28	2.69	0.15	0.79
Drumstick weight	227.25	221.75	228.42	221.08	222.42	0.78	2.19	0.53	0.99
Wing weight	174.17	171.83	174.83	162.92	166.83	0.20	1.87	0.08	0.87
Abdominal fat weight	48.77	46.43	44.81	48.77	45.72	0.76	1.14	0.66	0.68
Liver	46.82	45.01	45.80	43.76	45.71	0.68	0.65	0.47	0.41

3.3 Metabolizable energy

The effect of FFAs on metabolizable energy of 31 days-old broilers is shown in Table 4. The substitution of FFAs in 25, 50, 75 and 100% were not affected the metabolizable energy of broilers at 31 days of age. Vilà and Esteve-Garcia (1996a, b, c) [5-7], reported that the metabolizable energy was influenced by the unsaturated: saturated (U:S) ratio and that increased concentrations of saturated FFA decreased the metabolizable energy content, which in agreement with Wiseman et al. (1991) [25], who showed that increasing the proportion of FFAs while reducing the U:S ratio was associated with lower metabolizable energy, whereas increasing the FFAs with increase in the U:S ratio slightly improved the metabolizable energy. Therefore, the metabolizable energy in this study was increased due to the high ratio of U:S.

Table 4 Effect of free fatty acids supplementation in diet on metabolizable energy of 31 days old broilers.

Item	Free fatty acids					*P*-Value	SEM	Contrast, P<	
	0%	25%	50%	75%	100%			Linear	Quadratic
ME (Cal/kg)	3,110.17	3,135.08	3,184.77	3,141.53	3,167.84	0.11	9.54	<0.01	<0.01

4. Conclusion

In conclusion, FFAs substitution at 25%, 50%, 75% and 100% in diets had no negative effect to the body weight when compared to the control group. The carcass quality was not affected by FFAs. Moreover, there was no significant difference of the metabolizable energy in broilers fed with FFAs.

5. Acknowledgment

The authors gratefully acknowledge that this study was supported by Wawasan Agrolipids SDN BHD (Johor, Malaysia). Thank you to the staff from the Department of Animal Science, Kasetsart University, Thailand for suggestions, guidance and support throughout this trial.

References

1. H. Yacowitz, Supplementation of corn-soybean oil rations with penicillin and various fats, Poult. Sci. **32**, 930. (Abstr.) (1953)
2. M.L. Sunde, Use of animal fats in poultry feeds, I. Am. Oil Chem. Soc. **31** 49–52 (1954)
3. A.J. St. Angelo, Lipid oxidation in foods, Critical Review in Food Science and Nutrition. **36 (3)** 175-224 (1996)
4. R.Z. Zambiazi, Oxidation reactions of vegetable oils and fats, Revista da Sociedade Brasileira de Ciência e Tecnologia de Alimentos, Campinas. **33 (1)** 1-7 (1999)
5. B. Vilà, E. Esteve-Garcia, Studies of acid oil and fatty acids for chickens. (I) Influence of age, rate of inclusion and degree of saturation on fat digestibility and metabolizable energy of acid oils, Br. Poult. Sci. **37** 105–117 (1996a)
6. B. Vilà, E. Esteve-Garcia, Studies on acid oils and fatty acids for chickens. (II) Effect of free fatty acid content and degree of saturation of free fatty acids and neutral fat on fatty acid digestibility, Br. Poult. Sci. **37** 119–130 (1996b)
7. B. Vilà, E. Esteve-Garcia, Studies on acid oils and fatty acids for chickens. (III) Effect of chemical composition on metabolizable energy of by-products of vegetable oil refining, Br. Poult. Sci. **37** 131–144. (1996c)
8. J. Wiseman, F. Salvador, The influence of free fatty acid content and degree of saturation on the apparent metabolizable energy value of fats fed to broilers, Poult. Sci. **70** 573-582 (1990)
9. SAS Institute. SAS User's Guide: Statistics. Version 8 ed. (SAS Inst. Inc.,Cary, NC, 1996)
10. D.B. Duncan, Multiple range and multiple F tests, Biometrics, **11** 1–42 (1955)
11. G.M. Pesti, R.I. Bakalli, M. Qiao, K.G. Sterling, A comparison of eight grades of fat as broiler feed ingredients, Poult. Sci. **81** 382–390 (2002)

12. H. Wu, L.M. Gong, L. Guo, L.Y. Zhang, J.T. Li, Effects of the free fatty acid content in yellow grease on performance, carcass characteristics, and serum lipids in broilers, Poult. Sci. **90** 1992–1998 (2011)
13. D.W.F. Shannon, The effect of level of intake and free fatty acid content on the metabolizable energy value and net absorption of tallow by the laying hen, J. Agric. Sci. (Camb.). **75** 217– 221 (1979)
14. D.Sklan, Digestion and absorption of lipids in chicks fed triglycerides or free fatty acids: Synthesis of monoglycerides in the intestine, Poult. Sci. **58** 885–889 (1979)
15. J.B. Rossell, Measurement of rancidity, Pages 22–53 in Rancidity in Foods. 3rd ed. J.C. Allen and R.J. Hamilton, ed. (Chapman and Hall, London, UK., 1994)
16. C.C. Whitehead, C. Fisher, The utilization of various fats by turkeys of different ages, Br. Poult. Sci. **16** 481–485 (1975)
17. J. Wiseman, F. Salvador, The effect of age and rate of inclusion on the apparent metaboliszble energy values and apparent fat availability, Br. Poult. Sci. **30** 653–662 (1989)
18. R. Renner, F.W. Hill, Utilization of fatty acids by the chicken, J. Nutr. **74** 259–264 (1961a)
19. R. Renner, F.W. Hill, Factors affecting the absorbability of saturated fatty acids in the chicks, J. Nutr. **74** 254–258. (1961b)
20. R.J. Young, The energy value of fats and fatty acids for chicks. I. Metabolizable energy, Poult. Sci. **40** 1225–1233. (1961)
21. R.J. Young, R.L. Garrett, Effect of oleic and linoleic acids on the absorption of saturated fatty acids in the chicks, J. Nutr. **81** 321–329 (1963)
22. R.L. Garrett, R.J. Young, Effect of micelle formation on the absorption of neutral and fatty acids by the chicken, J. Nutr. **105** 827–838 (1975)
23. E. Ketels, G.D. Groote, Effect of ratio of unsaturated to saturated fatty acids of the dietary lipid fraction on utilization and metabolizable energy of added fats in young chicks, Poult. Sci. **68** 1506–1512 (1989)
24. E.D. Peebles, S.M. Doyle, T. Pansky, P.D. Gerard, M.A. Latour, C.R. Boyle, T.W. Smith, Effect of breeder age and dietary fat on subsequent broiler performance and slaughter yield, Poult. Sci. **78** 512–515 (1999)
25. J. Wiseman, F. Salvador, J. Craigon., Prediction of the apparent metabolizable energy content of fats fed to broiler chicks, Poult. Sci. **70** 1527–1533 (1991)

Forecasting Duck Egg Production Using ARIMA Model: A Case Study on Magelang Ducks

Zakia Ulfah[1,3], *Dyah* Maharani[1*], and *Heru* Sasongko[2]

[1]Laboratory of Animal Genetics and Breeding, Department of Breeding and Reproduction, Faculty of Animal Science, Universitas Gadjah Mada (UGM), Yogyakarta, Indonesia
[2]Poultry Science Laboratory, Animal Production Department, Faculty of Animal Science, Universitas Gadjah Mada (UGM), Yogyakarta, Indonesia
[3]Animal Breeding Division, Animal and Fisheries Department, Magelang Regency Government, Central Java, Indonesia

Abstract. This study investigated the application of the ARIMA model to forecast egg production trends in the Magelang ducks. Data were collected from 100 ducks over a 135-day period135 d of monitoring. The mean Daily Duck Production (DDP) was 43.53% with a standard deviation of 23.82%, indicating substantial variability in age-related traits and genetic factors. Following confirmation of data stationarity, the ARIMA (2,1,0) model was identified as the optimal fit. The model effectively captures historical trends and provides accurate forecasts of future production. The mathematic model was $Y_t = 0,896 + 0,322\ Y_{t-1} + 0,315\ Y_{t-2} - 0,363\ Y_{t-3} + \varepsilon_t$. These findings provide valuable insights into improving farm management, optimizing resource allocation, and enhancing egg production strategies. Thus, the ARIMA model can be used by researchers or farm managers to help farmers optimize resource management. This model helps farmers optimize resource management during peak and low production periods, making it highly useful even with limited data. Broadly, it enables better decision-making regarding feeding, breeding, and resource allocation for sustainable and profitable farm management.

1 Introduction

1.1 Background

Duck eggs occupy a significant and expanding position in poultry industry. In certain Asian regions, including China, duck eggs constitute a substantial share of total egg consumption, representing between 10% and 30% of the market[1]. Their nutritional benefits, culinary versatility, and increasing market demand underscore their potential as valuable products for producers and consumers. Forecasting egg production is crucial for enhancing poultry operations, facilitating prompt decision-making, and boosting efficiency through precise forecasts derived from past data and model assessments.[2]. This practice ultimately leads to improved profitability and long-term viability in the duck egg industry [3]. Furthermore, understanding future production trends assists farmers in aligning their marketing strategies

* Corresponding author: d.maharani@mail.ugm.ac.id

with consumer demand, ensuring that they can capitalize on peak seasons, even with limited data availability for the Magelang duck. Farmers generally face limitations in data availability owing to insufficient recording. Forecasting helps farmers manage risks by allowing them to anticipate production changes and to take proactive steps to address potential challenges. Effective forecasting enhances productivity and contributes to sustainability and profitability of poultry farming.

The autoregressive integrated moving average (ARIMA) method is a statistical tool for predicting egg production, particularly when dealing with time-series data that exhibit seasonal trends. By leveraging past observations, ARIMA models effectively identify patterns and autocorrelations in time-series data, thereby facilitating accurate prediction of future output levels[4]. The ARIMA model predicts future data points by understanding the autoregressive (p), degree of differencing (d), and moving average (q) of time-series data[5]. ARIMA forecasting of egg production utilizes a time-series analysis to predict future outputs based on historical data. The model enhances its forecasting accuracy by utilizing differencing techniques to handle non-stationary data, which is crucial for accurately predicting fluctuating egg production and similar applications[6]. Combining simplicity with predictive capabilities, ARIMA models offer an efficient approach to forecasting [7]. ARIMA is optimal for predicting Daily Duck Production (DDP) in the absence or minimal presence of seasonal patterns, offering simplicity, flexibility with non-stationary data, and robust short-term forecasts. Daily duck production refers to the number of eggs that a female duck lays per day. Daily duck production was calculated by expressing the average number of eggs laid per day relative to the number of female ducks in a flock. This is referred to as laying percentage or duck day production (DDP). Although SARIMA is suitable for pronounced seasonality, ARIMA's lower complexity renders it ideal for straightforward trend-based predictions.

1.2 Research Objectives

The objective of this study was to investigate the most suitable ARIMA model for forecasting Magelang ducks's egg production trends, considering the time-series characteristics of DDP data on Magelang duck. This study proposed that the ARIMA model is well suited for predicting egg production trends, given the time-series characteristics of Daily Duck Production (DDP) data.

2 Materials and Method

2.1 Study Design

Thus data collection was performed by completely randomized design. Seventy days aged female Magelang ducks were randomly selected based on hatching records to ensure accurate age documentation. The selection included 100 ducks that were in good health, without physical defects, and reared under the same conditions as individual cages. The number of eggs collected daily was recorded to as a critical metric to determine DDP for assessing the farm productivity.

The first day of recording commenced with discovery of the first egg in the cage. Observations were recorded daily for 135 days after the first egg was laid, number of eggs daily data were recorded by KTT Bina Mandiri, Magelang, and Central Java between April 27, 2024 (157 aged old), and September 8, 2024 (291 aged old).

2.2. Data Analysis

Duck Day Production (DDP) is a quantitative measure of egg-laying efficiency in duck flocks. It is calculated by comparing the number of eggs daily to the number of ducks observed, then multiplying by 100 to yield a percentage. This calculation serves as an effective metric for assessing the productivity of egg-laying ducks in a given population. The DDP data were then analyzed using SPSS 26 to highlight key aspects such as mean, median, standard deviation, frequency distribution, and coefficient variation, which were all examined through descriptive statistics. A DDP data trend analysis was performed to identify the patterns and determine the stationarity of the time series. Finally, an Autoregressive Integrated Moving Average (ARIMA) analysis was conducted. Additionally, using DDP as a time series, this study demonstrates the accuracy of the ARIMA model in forecasting future production, aiding in informed management decisions to optimize efficiency and resource allocation on duck farms.

3 Result and Discussion

3.1 Descriptive Statistic

Table 1. describes the descriptive statistics of the DDP data for Magelang ducks. A 78 percentage-point range indicates substantial variation in egg production percentages, presumably influenced by genetic variation among the ducks and age-related traits since the data were time series.

Table 1. Descriptive statistics of Magelang ducks Daily Duck Production (DDP)

No	Metric	Value
1	Number of duck (N)	100
2	Number of Days	135
3	DDP Range	78%
4	DDP Minimum	1%
5	DDP Maximum	79%
6	DDP Mean	43.53%
7	DDP Standard Deviation	23.82%
8	DDP Coefficient of Variance	54.72%

An average DDP of 43.53% reflected moderate productivity, with data collected when the ducks were 291 days old. Duck egg production depends on breed, age, environment, and health, with DDP ranging from 60.57% to 68.86% in various breeds over 12 weeks[8]. Research on Chinese Shan Ma ducks showed peak egg production at 180 days, surpassing 95%, and then declining to 40-50% by 520 days[9]. The high standard deviation of 23.82% suggests considerable daily variability in egg production, potentially owing to age-related trends and genetic variation. The standard deviation shows daily changes in egg production, which makes sense since ducks are aging. As they grew older, they produced more eggs, and genetic variation could also contribute to this variance, which represents a high coefficient of variance value of 54.72%.

3.2 Forecasting egg production by ARIMA model

ARIMA is a forecasting technique for time-series data that can handle trends and nonseasonality. The ARIMA model is consisted of (p, d, q) where, p is the autoregressive parameters, d is the number of differencing passes and q is the moving average parameters. Four critical steps must be considered when applying ARIMA: identification, estimation, diagnostic verification, and forecasting.

3.2.1. *Identification*

The process of achieving stationarity and constructing a suitable model involved an identification phase using autocorrelation functions (ACFs), partial autocorrelation functions (PACFs), and transformations such as differencing and logarithmic conversion [3]. The DDP data qualify as time-series data, making the ARIMA analysis applicable for forecasting. Fig 1 illustrates the Magelang Duck DDP trend over 135 days. The initial DDP blue graph is nonstationary, which necessitates data differencing [10]. After differencing, the red graph shows stationarity. Consequently, the d value in the ARIMA model is 1, as a Single-differencing was performed.

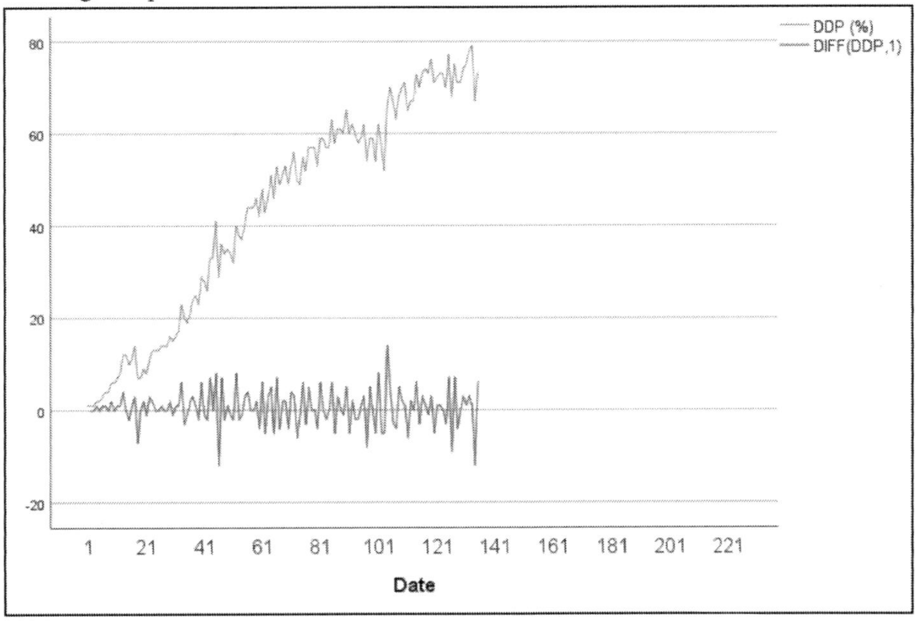

Fig. 1. The Magelang duck production and differencing trend over 135 days (DDP: duck daily production; diff = 1 ; N = 100)

Subsequently, we performed autocorrelation and partial autocorrelation analyses of the differenced DDP data.

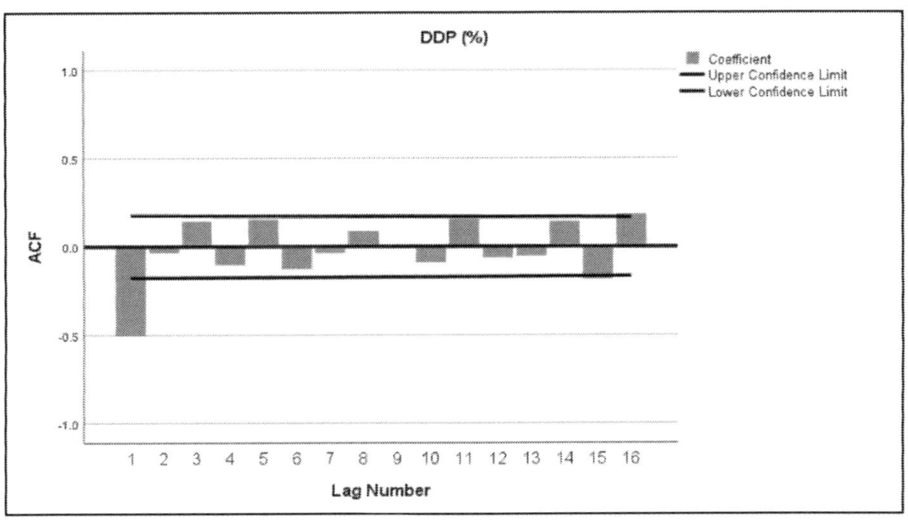

Fig. 2. Autocorrelation (ACF) chart of Magelang duck daily production (DDP)

The autocorrelation function (ACF) plot is shown in Fig. 2. Thus, potential q-value was determined. Because the lag exceeding the upper confidence limit is lag 1, the q value may be 0 or 1.

The p-value can be determined by analysing the partial autocorrelation function (PACF) chart. The lags exceeding the upper confidence limit are Lag 1 and Lag 2; consequently, the p-values applicable to the model are 0, 1, and 2. The autocorrelation function (ACF) and PACF charts indicate that the data are stationary, which is a prerequisite for implementing the ARIMA model. Partial autocorrelation is illustrated in Fig. 3.

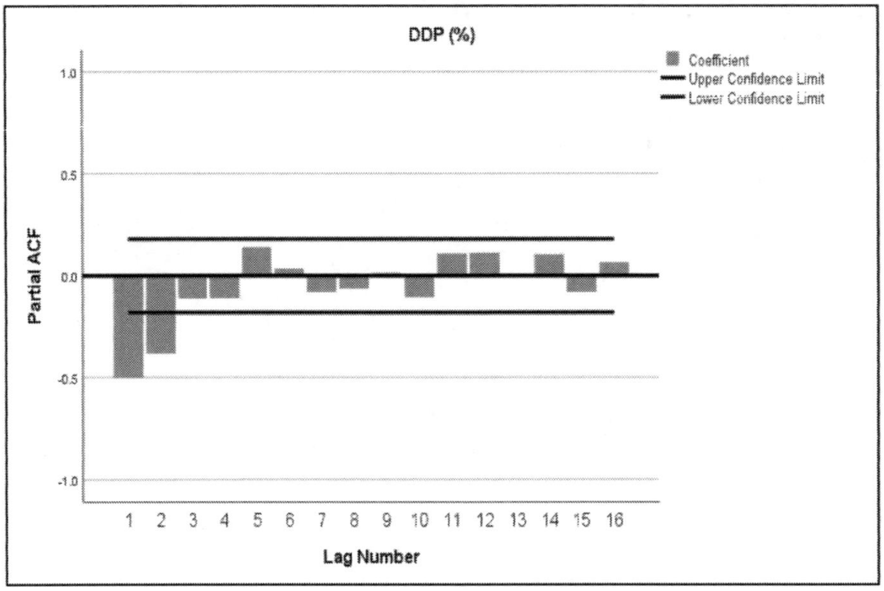

Fig. 3. Partial Autocorrelation (PAFC) chart of Magelang duck daily production (DDP)

Lags exceeding the upper confidence limit were lag 1 and lag 2. Consequently, we defined the potential p-values for the model as 0, 1, and 2. Based on our identification process, we can now determine the potential order of the ARIMA model that we will estimate in the next step. The potential values for p were 0, 1, and 2, d was set to 1, and q was either 0 or 1. The The ARIMA model's ordo suggested for forecasting egg production is presented in Table 2.

Table 2. The ARIMA model's ordo suggested for forecasting egg production

No	p	d	q
1	0	1	0
2	1	1	0
3	2	1	0
4	0	1	1
5	1	1	1
6	2	1	1

3.2.2. *Estimation*

The ARIMA models for predicting egg production were fitted using Statistical Package for Social Sciences (SPSS) software to estimate the model parameters. The estimation or specification step calculates the coefficients based on least-squares and likelihood methods, including the BIC value. The prediction of Magelang duck egg production using ARIMA Model analysis results are presented in Tables 3 and 4.

Table 3. The prediction of Magelang Duck egg production using The ARIMA Model Analysis (R square and error)

No	ARIMA	R square	Error			
			RMSE	MAPE	MAE	BIC
1	ARIMA (0,1,0)	0.97	4.06	10.09	3,02	2.88
2	ARIMA (1,1,0)	0.98	3.52	9.83	2.61	2.63
3	ARIMA (2,1,0)	0.98	3.31	9.75	2.47	2.54
4	ARIMA (0,1,1)	0.98	3.27	9.97	2.49	2.58
5	ARIMA (1,1,1)	0.98	3.27	10.03	2.49	2.52
6	ARIMA (2,1,1)	0.98	3.27	10.03	2.49	2.55

The selection of the optimal ARIMA model involved the evaluation of various configurations, with the ARIMA (2,1,0) model exhibiting superior performance. ARIMA (2,1,0) achieved a high R^2 value of 0.98, indicating that it accounts for 98% of the variation in egg production, thus capturing underlying trends effectively. It also exhibited low error metrics: RMSE of 3.31, MAPE of 9.75, and MAE of 2.47, making it more accurate than the other models. Additionally, it had the lowest BIC score of 2.54, indicating a good balance between model complexity and fit [7]. These low error metrics and high R^2 values collectively demonstrate the model's ability to minimize errors and produce accurate forecasts, reflecting its effectiveness in capturing data variability and generating reliable predictions with minimal inconsistencies [11].

Table 4. The prediction of Magelang duck egg production msing The ARIMA model analysis (variables and white noise assumption)

No	ARIMA	Variables		White Noise Assumption (LJung Box Q test)	
		Significant	Unsignificant	P value	White noise
1	ARIMA (0,1,0)	0	2	0.00	No
2	ARIMA (1,1,0)	2	0	0.06	Yes
3	ARIMA (2,1,0)	3	0	0.70	Yes
4	ARIMA (0,1,1)	2	0	0.85	Yes
5	ARIMA (1,1,1)	2	1	0.84	Yes
6	ARIMA (2,1,1)	2	2	0.85	Yes

As shown in Table 4, the Ljung-Box Q-test for white noise gives a p-value of 0.70, which is relatively high, indicating that the residuals resemble white noise and the model is well-specified. Among the various models, ARIMA (2,1,0) stands out with the greatest number of significant parameters, totalling 3. Thus, ARIMA (2,1,0) offers a more precise and reliable forecast than other ARIMA models, making it the best choice for analysis. Overall, ARIMA (2, 1, 0) is a reliable tool for forecasting egg production trends, providing insights for farm management and resource optimization. This information is vital for researchers and analysts assessing model reliability and effectiveness in predictions.

3.2.3. *Diagnostic*

The diagnostic process involves analysing parameter significance through visual representations and statistical measures to assess model fit [11]. Table 4 demonstrates that the ARIMA (2,1,0) model is optimal due to its incorporation of the highest number of statistically significant parameters compared to alternative models. Non-significant parameters were excluded to enhance the model's reliability and ensure that only the variables meaningfully contributing to the model's predictive capability were retained.

The model visual representation shown in Fig. 4 provides a reasonably good fit to the observed data, and the forecast shows a steady trend with some uncertainty, reflected in the widening confidence intervals. This is typical for ARIMA or similar time-series models, where future predictions are based on past trends and uncertainty grows as we move further from the last observed point. The area between the upper Control Limit (UCL) and lower Control Limit (LCL) reflects the range of uncertainty or variability in the forecast. The forecasted data stays within the UCL and LCL, it suggests that the model is performing as expected. Forecasting process with the model (2,1,0) for forecasting of future egg production indicated a good fitting of the model for prediction.

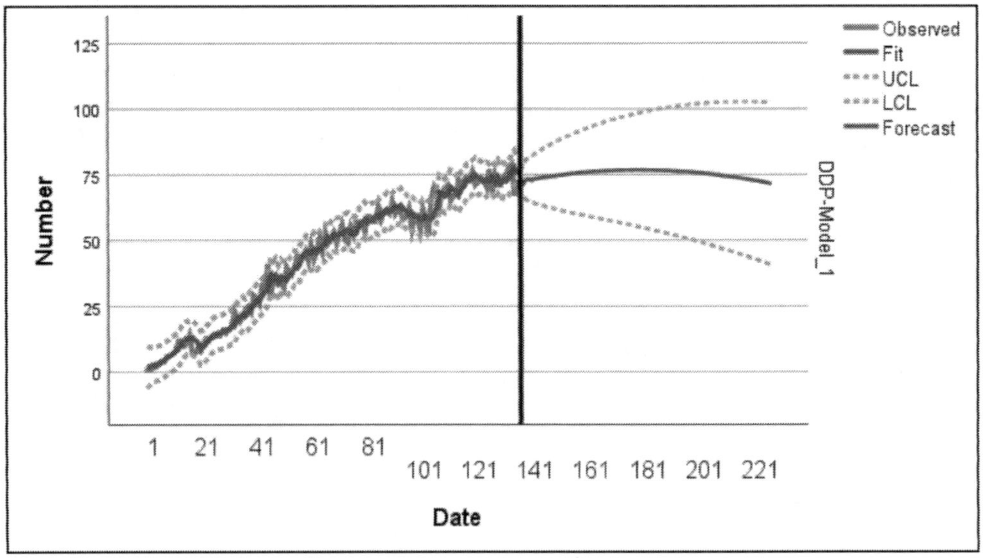

Fig 4. Actual and estimated egg production of Mageland duck based on ARIMA (2,1,0)

3.2.4. *Forecasting*

Based on analysing parameter significance through visual representations and statistical measures, we determined that the ARIMA (2,1,0) mathematical model is the most suitable for forecasting egg production in Magelang ducks. The significant AR terms (lag 1 and lag 2), as shown in Table 5, indicate that the past values of the series strongly influence future values in a negative direction. The predictive capability of the model is significantly bolstered by the presence of the constant term. These parameters confirm that the ARIMA (2,1,0) model is well-fitted and reliable for forecasting, as all components are statistically significant at a high confidence level.

Table 5. The Prediction of Magelang Duck Egg Production Using The ARIMA (2,1,0) Model Parameters

ARIMA (2,1,0) Model Parameters								
					Estimate	SE	t	Sig.
DDP (%)-Model (2,1,0)	DDP (%)	No Transformation	Constant		.869	.288	3.014	.003
			AR	Lag 1	-.678	.082	-8.263	.000
				Lag 2	-.363	.085	-4.272	.000
			Difference		1			

In general, the mathematical model based on ARIMA (2,1,0) is as follows::

$$Y_t = \underbrace{\Phi_0 + \Phi_1 Y_{t-1} + \Phi_2 Y_{t-2} \ldots + \Phi_n Y_{t-n}}_{AR} + \underbrace{\varepsilon_t - \theta_1\varepsilon_{t-1} - \theta_2\varepsilon_{t-2} - \ldots \theta_q\varepsilon_{t-q}}_{MA} \quad (1)$$

The parameter analysis results obtained using SPSS 26, presented in Table 43, were incorporated into the formula. Given that the model is differenced once, as indicated by the d value of 1, it is necessary to reformulate it in terms of the original *Yt* values. The P value of the constant was < 0.05 (0.003) and this indicated that the model fitted with the constant The mathematical model of ARIMA (2,1,0) is as follows:

$$\Delta Y_t = 0{,}896 - 0{,}678(Y_{t-1} - Y_{t-2}) - 0{,}363\,(Y_{t-2} - Y_{t-3}) + \varepsilon_t \quad (2)$$
$$Y_t - Y_{t-1} = 0{,}896 - 0{,}678(Y_{t-1} - Y_{t-2}) - 0{,}363\,(Y_{t-2} - Y_{t-3}) + \varepsilon_t \quad (3)$$
$$Y_t = 0{,}896 + Y_{t-1} - 0{,}678\,Y_{t-1} + 0{,}678\,Y_{t-2} - 0{,}363 Y_{t-2} + 0{,}363\,Y_{t-3} + \varepsilon_t \quad (4)$$

So the final mathematic model was :

$$Y_t = 0{,}896 + (1-0{,}678)\,Y_{t-1} + (0{,}678-0{,}363)\,Y_{t-2} - 0{,}363 Y_{t-3} + \varepsilon_t \quad (5)$$
$$Y_t = 0{,}896 + 0{,}322\,Y_{t-1} + 0{,}315\,Y_{t-2} - 0{,}363\,Y_{t-3} + \varepsilon_t \quad (6)$$

Where:
Y_t : The value of Daily Duck Production (DDP) at time t
Y_{t-1} : This is the DDP at time t-1 (the previous day)
Y_{t-2} : This is the DDP at time t-2 (two days ago)
Y_{t-3} : This is the DDP at time t-3 (three days ago)
ϵ_t : The error term at time t.

Based on the mathematical model, it appears that peak egg production was likely reached around day 161 after the first egg laying recorded refers to week 23, with a value of 77%. The previous research indicated that the peak of egg production of Magelang ducks was 60,57% lower than that of Mojosari ducks were 68.86% [8]. This means that the peak production was higher in this study, which may have been caused by intensive rearing using individual cages. A previous study also [12] revealed that egg production in Alabio ducks occurs at week 24. This means that Magelang ducks are predicted to reach egg production a week earlier than Alabio ducks. The value at peak production is 77%, which is lower than that in the Shan Ma duck, which exceeds 90% and reaches 180 days old duck [9]. These results suggest that farmers must enhance the nutrition and management of their farms to optimize egg production. Duck egg production can be enhanced through various strategies, including dietary modifications[13], genetic improvements[14], and management practices[15, 16]. The ARIMA (2, 1, 0) model predicts future values based on past observations by applying first differencing (d=1) for stationarity and using two lagged observations (p=2) without a moving average component (q=0). While suitable for forecasting egg production, advanced methods, such as ETS or SARIMA, are preferred when seasonal patterns or external factors are significant for the next study. ARIMA's simplicity and reliability of ARIMA make it appropriate for datasets lacking strong seasonal variations. Egg production can be forecasted using limited data, such as a single time series, helping duck farmers manage real-world data limitations caused by poor recording. This tool helps farmers optimize resource management during peak and low production, proving useful even with limited data. Broadly speaking, it improves decisions on feeding, breeding, and resource allocation for sustainable and profitable farming.

4 Conclusion

The ARIMA (2,1,0) model was identified as the optimal fit, effectively capturing historical trends and providing accurate forecasts of future egg ducks production. These findings provide valuable insights into improving farm management, optimizing resource allocation, and enhancing egg production strategies. Thus ARIMA model can conducted by researcher or farm manager to help farmers optimize resource management during peak and low production periods, making it highly useful, even with limited data. Predictive models, such as ARIMA, enhance resource efficiency and sustainability, allowing farmers to adjust operations according to predicted trends, thereby reducing environmental impacts and increasing profitability. The ARIMA model provides valuable insights for decision-making regarding future egg production of Magelang ducks. Broadly, it enables better decision-making regarding feeding, breeding, and resource allocation for sustainable and profitable farm management. This forecasting technique is not only beneficial for this particular farm but can also be effectively utilized by ducks egg producers worldwide to forecast future performance.

This research was funded by the LPDP Scholarship, Ministry of Finance of the Republic of Indonesia, Faculty of Animal Science at UGM, and the Animal and Fisheries Department of Magelang Regency. All errors and omissions were responsible for this study. This study was approved by the Ethics Committee of the Faculty of Veterinary Medicine, UGM (ethical clearance no. 144/EC-FKH/eks./2024, issued on January 8, 2024). Funding was provided by the LPDP (Scholarship No. 202212210312291), with no involvement in the study design, data collection, analysis, or manuscript preparation. We sincerely thank Mr. Bowo, Chairman of KTT Bina Mandiri, Kalangan Ambartawang, Mungkid, Magelang, for his invaluable support in providing the research site and assisting with data collection. The data supporting this study are available from the corresponding author upon request.

Author contributions: Z. Ulfah: Study conceptualization, data collection, analysis, interpretation, and manuscript writing. D. Maharani: intellectual support and manuscript revision. H. Sasongko: Experimental design and manuscript revision.

References

1. J. Sumarmono, Duck production for food security, The 1st Animal Science and Food Technology Conference (AnSTC) in IOP Conference Series. J. Earth and Env. Sci., **372**, 012070. (2019)

2. N. Bumanis, et al., Hen Egg production forecasting: capabilities of machine learning models in scenarios with limited datasets. J. App. Sci. **13**, 7607 (2023).
3. A. Wijaya, et al. Predictions of duck egg production in 34 Indonesian provinces using backpropagation algorithm. International Conference on Informatics, Multimedia, Cyber and Informations System (ICIMCIS), Jakarta, Indonesia, 7-8 November 2023, Desember, 14 (2023), 27-32.
4. E.B. Sinu, M.A. Kleden, and A. Atti, Application of ARIMA model for forecasting national economic growth: a focus on gross domestic product data, Berekeng, J. Math. Sci. App., **18,** 1261-1272 (2024).
5. C. Deemee, et al. Statistical comparison ARIMA order performance in stock market. in 2023 7th International Conference on Information Technology (InCIT). Chiang rai, Thailand, 16-17 November 2023, January, 30 (2024), 81-85.
6. D.K Yadav, K. Soumya, and L. Goswami, Autoregressive integrated moving average model for time series analysis. International Conference on Optimization Computing and Wireless Communication (ICOCWC), Debre Tabor, Ethiopia, 29-30 January 2024, March, 21 (2024), 1-6.
7. Y. Zhang and G. Meng. Simulation of an adaptive model based on AIC and BIC ARIMA predictions. J. Phy, **2449**, 1-7 (2023).
8. H. Henrik and M. Marhayani, Egg production and quality of Magelang duck, Mojosari duck, and their reciprocal crosses. Jurnal Ilmu-Ilmu Peternakan, **30**, 180-83, (2020)
9. Z. Zhu, et al., Ovarian transcriptomic analysis of Shan Ma ducks at peak and late stages of egg production, Asian-Australian J. Anim.Sci., **30**, 1215-1224 (2017).
10. R.J. Hyndman and G. Athanasopoulos, 8.1. Stationarity and differencing. Forecasting: Principles and practices, (O texts, Melbourne, Australia, 2013).
11. F.D.M Abdallah, Role of time series analysis in forecasting egg production depending on ARIMA model. J.App. Math., **9**, 1-5 (2019).
12. T. Susanti and L. Prasetyo. Pendugaan parameter genetik sifat-sifat produksi telur itik alabio. J. Tech. Anim. Vet. dan Veteriner, (2008)
13. J. Loing, et al., Increasing duck egg production using grow enhancer treatment. International J.1 of Bio. Sci., **6,** 31-35 (2015).
14. M. Oyebanjo, O. Osaiyuwu, and A. Salako, Prolactin, genetics, and bioinformatics: The trinity for improving duck egg production in Nigeria–A review. J. Anim. Sci. and Vet. Med., **8**(5), 235-246 (2023).
15. H. Sasongko, Feeding management evaluation of duck farmer groups in Brebes. in International Seminar on Tropical Animal Production (ISTAP). 2010. Yogyakarta.
16. I. Arhipova, et al. Smart platform designed to improve poultry productivity and reduce greenhouse gas emissions. in Proceedings of Sixth International Congress on Information and Communication Technology: ICICT 2021, Singapore, September, 24, (2021), 35-46.

Effect of a mixture of monoglycerides with salinomycin on the growth performance and intestine morphology in broiler chickens

Hutsanai Siriphong[1], *Sombat* Prasongsook[1], *Choawit* Rakangthong[1], *K.Teepalak* Rangubhet[1] and *Theerawit* Poeikhampha[1]*

[1]Department of Animal Science, Faculty of Agriculture, Kasetsart University, Bangkok, Thailand

Abstract. This study aims to interact with the combined influence of monoglyceride supplementation and salinomycin on growth performance and the morphology of the small intestine in broiler chickens. A factorial treatment design was employed in a completely randomized design (CRD). The broiler chickens were randomly divided into four experimental groups, the treatments studied were as follows: 1) Control group (without monoglyceride and salinomycin supplementation), 2) 0.05% monoglyceride supplementation combined without salinomycin, 3) 0.006% salinomycin without monoglyceride supplementation, and 4) 0.05% monoglyceride supplementation combined with 0.006% salinomycin. The study results indicated that the combined supplementation of monoglycerides and salinomycin had an interaction effect on the feed conversion ratio during the first 10 days, villus height: VH in the duodenum, and crypt depth: CD with villus height/crypt depth ratio: CD in the jejunum, which were significantly different statistically. Therefore, the supplementation of 0.05% monoglycerides combined with 0.006% salinomycin has an interaction effect on the feed conversion rate to body weight (days 1-10), the increase in VH in the duodenum, and the CD, along with the VH: CD ratio in the jejunum, resulting in improved growth performance and morphology of the proximal small intestine in broiler chickens.

1 Introduction

The poultry industry is a popular source of protein but is now facing the problem of cocidiosiss which causes damage to this industry. Coccidiosis affects productivity and intestinal health. However, the abuse of antibiotic additives induced transferable

* Corresponding author: agrtrw@ku.ac.th

antibiotic resistance in pathogenic bacteria in both birds and human beings.[3] Reducing or replacing the antibiotics in poultry feed has been accepted and promoted worldwide, which urges the scholars and enterprises to find alternative feed additives for the poultry industry.

Salinomycin, an ionophore coccidiostat, is commonly used as a supplement in poultry feed to manage infections caused by *Eimeria spp.* and to prevent necrotic enteritis induced by *Clostridium perfringens* in poultry effect on FCR[6].which aligns with research Conway broiler chickens fed 60 ppm salinomycin for 5 or 6 weeks followed by a withdrawal period of 1 or 2 weeks showed increased feed consumption during the withdrawal period and improved feed conversion ($P < 0.05$) in comparison with unmedicated birds.[2] However, the trend of the poultry market has been changing. For example, American farms raise 60% poultries without using feed containing antibacterial drug. [1] Therefore, preventing cocidiosis without using antibacterial drug in feed is a raising trend and emphasizing the importance of vaccines or mixture of monoglycerides for achieving better effective production is a topic that requires extensive study in the poultry industry.[9]

Combining monoglycerides of butyric-, capric-, and caprylic acid or monoglycerides blend which can be absorbed in small intestine [12] which aligns with the research of Patcharanan et al. [10] who studied the effect of monobutyrin in feed and found that it helps increasing egg production, improving egg quality, and increasing the height of villi in the small intestine which would improve the effectiveness of digesting and absorbing system of poultry after the highest performance of production. Monoglycerides and butyric acid can be absorbed directly in the mid and distal parts of the small intestine. Also, monobutyrin can be broken down into butyric acid in the intestinal mucosa. [15] which aligns with the research of Lin et al. supplement with monoglycerides showed beneficial efficacy on the growth and intestinal function of broilers, demonstrating the potential value of monoglycerides to poultry industry. In terms of dosage, the current trial shows that 3,000 g/t (1−21 d) and 1,000 g/t (22−42 d) monoglycerides without chlortetracycline were the appropriate dietary supplemented rate for broilers. And the mixed use of 500 g/t chlortetracycline and 1,000 g/t monoglycerides was benefit for broilers at 22 to 42 d. [7]

The present study aimed to evaluate the efficacy of monoglycerides combined with salinomycin in broiler diets on growth performance and intestinal morphology in broiler chickens. We hypothesized that the interaction of the mixture of monoglycerides and salinomycin would have positive effects on intestinal histomorphology and improve performance.

2 Materials and methods

2.1 Experimental design and diets

The experiment utilized 1,260 one-day-old ROSS 308 broiler chickens, with a total rearing period of 35 days. A factorial treatment design was employed in a completely randomized design (CRD) for the experiment. All broiler chickens were randomly assigned to four experimental groups, each consisting of six replications with 35 chickens per experimental unit. The treatments under investigation are detailed in Table 1.

Table 1. The level of monoglyceride supplementation combined with salinomycin in broiler chicken feed.

Factor (a)	Factor (b)	Treatment combination
Level of SLM [2]	Level of MGB [3]	a_1b_1 = 0%SLM-0%MGB
a_1 = 0%	b_1 = 0%	a_1b_2 = 0%SLM-0.05%MGB
a_2 = 0.006%	b_2 = 0.05%	a_2b_1 = 0.006%SLM-0%MGB
		a_2b_2 = 0.006%SLM-0.05%MGB

[2]Level of SLM is salinomycin sodium 2 levels according to the FDA guidelines 21 CFR part 558 subpart B Specific New Animal Drugs for Use in Animal Feeds.
[3]Level of MGB is monoglycerides of butyric-, capric- and caprylic acids according to published research of Performance and gut health status of broiler fed diets supplemented with two graded levels of a monoglyceride blend [11]

The experimental feed used the main ingredients of corn and soybean meals, according to the nutritional requirements of ROSS 308 broiler chickens. The feed was divided into three phases: the starter phase (1-10 days), the grower phase (11-24 days), and the finisher phase (25-35 days) (Tables 2 and 3). Throughout the experiment, the broiler chickens had *ad libitum* access to feed and water. The withdrawal period (without salinomycin) lasted for the last 7 days of the experiment.

2.2 Housing management

The experiment was conducted at the Suwannawajakasikij Animal Research Farm, Department of Animal Science, Faculty of Agriculture, Kasetsart University, Bangkok. The number of approval for Animal Care and Use for Scientific Research Kasetsart University is ACKU65-AGR-016. Raising broiler chickens in a floor-released system within a closed housing facility, equipped with an environmental control system utilizing an evaporative cooling system. There are four 50-inch exhaust fans installed at the end of the facility, which operates under a tunnel ventilation system. The pens are partitioned with dimensions of 1.5 meters wide, 3 meters long, and 0.5 meters high, totaling 36 pens. The pen floors are made of concrete and covered with approximately 5 centimeters of rice husk as bedding material. During the first 10 days of the experiment, heat lamps for the chicks were installed in each pen, and feeding is done using yellow trays designed for chicks, while water was provided using chick waterers. From days 11 to 35, feeding was conducted using hanging feeders with a diameter of 10 inches, and an automatic nipple watering system with 7 nipples per pen was implemented.

Table 2. Demonstrate raw materials ratio of each type of feed ingredient in the diets of broiler chickens at different stages.

Ingredients (%)	**Starter**	**Grower**	**Finisher**
Corn	55.10	41.62	40.99
Wheat	-	10.00	15.00
Rice bran defatted	-	5.00	5.00
Corn DDGs	-	8.00	8.00
Soybean oil	2.78	4.97	5.99
Soybean meal 48%	37.63	21.13	15.94
Rape Seed meal	-	4.00	4.00
Feather meal	-	2.00	2.00
L-Lysine hydrochloride	0.19	0.35	0.35
DL-Methionine	0.32	0.24	0.21
L-Threonine	0.09	0.08	0.06
Salt	0.35	0.35	0.35
Calcium carbonate	1.21	1.02	1.08
Mono dicalcium phosphate 21%	1.80	0.73	0.53
Vitamin and Mineral Premix*	0.24	0.24	0.24
Choline chloride 60%	0.28	0.27	0.25
Phytase and Xylanase	-	0.01	0.01
Total	100.00	100.00	100.00

*The vitamin and mineral premix consists of minerals such as copper, iodine, iron, manganese, selenium, and zinc, as well as vitamins including vitamin A, vitamin D, vitamin E, vitamin K, thiamine, riboflavin, niacin, pantothenic acid, pyridoxine, biotin, folic acid, and vitamin B12.

2.3 Growth performance measurements

On days 1, 10, 24, and 35 of the experiment measure according to ROSS 308 guideline, individual body weight of chickens and feed consumptions were weighed using a digital scale (OHAUS Valor 1000 series) to record the initial body weight (IBW) and final weight (FW). This data was used to calculate body weight gain (BWG) and average daily gain (ADG), feed intake (FI), feed conversion ratio (FCR), and mortality (M).

Table 3. Nutrient composition requirements of broilers at starter, grower and finisher

Nutrient	**Starter**	**Grower**	**Finisher**	**Units**
Metabolize energy for Poultry	3,000	3,100	3,200	Kcal/kg
Protein	23.00	21.50	19.50	%
Fat	5.27	7.81	8.82	%
Fiber	2.56	3.58	3.47	%

Calcium	0.96	0.81	0.78	%
Total Phosphorus	0.72	0.58	0.52	%
Available Phosphorus for Poultry	0.48	0.44	0.39	%
Salt	0.34	0.34	0.34	%
Digestible of amino acids				
Arginine	1.44	1.22	1.08	%
Histidine	0.58	0.50	0.45	%
Isoleucine	0.90	0.79	0.70	%
Leucine	1.81	1.67	1.53	%
Lysine	1.28	1.15	1.02	%
Methionine	0.63	0.53	0.48	%
Methionine and Cysteine	0.95	0.87	0.80	%
Phenylalanine	1.05	0.95	0.86	%
Phenylalanine and Tyrptophan	1.77	1.36	1.20	%
Threonine	0.86	0.77	0.68	%
Tryptophan	0.25	0.21	0.19	%
Valine	0.98	0.92	0.83	%
Choline	1,700	1,600	1,500	mg/kg.

2.4 Intestinal morphology measurements

Samples were collected (one bird representative individual per pen, with six repetitions per treatment) from three section: the duodenum, jejunum, and ileum according to the method of Srinual, Punyatong et al. [14] The samples were processed using an automatic tissue processor, followed by tissue embedding with a tissue embedding machine, and then sectioned using a microtome. The sections were stained with hematoxylin and eosin, and the morphology of the small intestine was recorded using a light microscope at 40x magnification in conjunction with a digital camera. Morphological data of the small intestine were analyzed, including villus height (VH) and crypt depth (CD). The details of the morphological analysis of the small intestine are as follows: villus height was measured from the tip of the villus to the base of the villus, excluding the portion of the measurement for crypt depth, which was measured from the basement membrane to the crypt mouth. The ratio of villus height to crypt depth (VH:CD) was then calculated.
The muscularis mucosae layer, which is thin and located adjacent to the lamina propria, consists of clearly visible smooth muscles and was measured from the upper part of the crypt base.

2.5 Statistical analysis

Data from the growth performance and intestinal morphology recordings were analyzed for variance using a factorial experimental design in a completely randomized trial (factorial experiments in CRD). Data were compared for differences between treatment means using Duncan's multiple range test. The differences between means were set at a significance level of $P < 0.05$ using the SAS® University Edition (Durner 2021) computer program using the following statistical model. [5]

3. Results

3.1 Growth Performance

The supplementation of monoglycerides or salinomycin significantly affected final weight (at 10, 24, and 35 days of age), growth rate (during the ages of 11-24 and 1-35 days), weight gain (during the ages of 1-10, 11-24, and 1-35 days), and feed conversion ratio (during the ages of 1-35 days) with significant differences (p<0.01). The supplementation of salinomycin also had a significant improved on weight gain (during the ages of 25-35 days), feed conversion ratio (during the ages of 11-24 days), and growth rate (during the ages of 1-10 and 25-35 days), with statistically significant differences (p<0.01) as shown in Tables 4 and 5.The supplementation of monoglycerides in conjunction with salinomycin has a significant joint effect on the feed conversion rate to body weight during the period of 1-10 days, with a statistically significant difference (p=0.03)

3.2 Intestinal Morphology

Villus heights in the duodenum were significantly increased in chickens fed with MGBxSLM compared to the control group or those receiving MGB or SLM alone. Additionally, the supplementation of SLM tended to increase crypt depth and the villus-to-crypt ratio (VH:CD). Villus heights in the Jejunum were also enhanced by SLM supplementation. Furthermore, supplementation with MGB, SLM, and the interaction between these groups significantly increased crypt depth and VH:CD in the jejunum. Additionally, supplementation with MGB and SLM increased villus heights in the ileum of broiler intestines, as shown in Table 6.

Table 4. Effect of monoglyceride with salinomycin interaction on growth performance in broiler chickens

Items	MGB		SLM		SEM[1]	*P*-value		
	0.00%	0.05%	0.00%	0.006%		MGB[2]	SLM[3]	MGB ×SLM[4]
Initial Body Weight (g)								
1 d	42.66	42.65	42.66	42.64	0.06	0.91	0.94	0.96
11 d	346.14	356.68	346.59	356.24	1.93	0.005	0.0012	0.29
25 d	1413.80	1454.68	1400.56	1467.92	9.51	0.0005	<0.0001	0.21
Final Body Weight (g)								
10 d	346.14	356.18	346.59	356.24	1.93	0.0005	0.0012	0.29
24 d	1413.80	1454.68	1400.56	1467.92	9.51	0.0005	<0.0001	0.21
35 d	2560.12	2615.31	2531.01	2644.42	16.39	0.01	<0.0001	0.36
Body Weight Gain (g)								
1-10 d	303.48	313.39	303.93	312.94	1.89	0.0012	0.0027	0.20
11-24 d	1067.66	1098.00	1053.97	1111.68	8.16	0.0043	<0.0001	0.30
25-35 d	1146.31	1160.63	1130.45	1176.50	10.86	0.5	0.04	0.77
1-35 d	2517.46	2572.66	2488.35	2601.77	16.37	0.01	<0.0001	0.36

[1] Pooled standard error of the mean.
[2] monoglycerides 0.05% in diet
[3]salinomycin 0.006% in diet
[4]Interaction of monoglycerides and salinomycin.

Table 5. Effect of monoglyceride with salinomycin interaction on growth performance in broiler chickens

Items	MGB		SLM		SEM[1]	*P*-value		
	0.00%	0.05%	0.00%	0.006%		MGB[2]	SLM[3]	MGB× SLM[4]
Average Daily Gain (g/d)								
1-10 d	30.68	31.18	30.38	31.56	0.22	0.12	0.0027	0.06
11-24 d	76.26	78.43	75.28	79.41	0.58	0.0043	<0.0001	0.30
25-35 d	104.21	105.51	102.77	106.96	0.99	0.49	0.04	0.77
1-35 d	74.04	75.67	73.19	76.52	0.48	0.01	<0.01	0.36
Feed Conversion Ratio (g/g)								
1-10 d	1.08	1.06	1.10	1.03	0.01	0.31	0.01	0.03
11-24 d	1.37	1.33	1.39	1.31	0.01	0.06	0.0016	0.21
25-35 d	1.57	1.55	1.58	1.54	0.01	0.40	0.23	0.75
1-35 d	1.43	1.39	1.44	1.38	0.01	0.03	0.0012	0.95
Mortality (%)								
1-10 d	0.25	0.36	0.49	0.00	0.20	0.27	0.81	1.00
11-24 d	0.51	0.25	0.77	0.00	0.21	0.08	0.53	0.53
25-35 d	0.51	0.79	1.04	0.27	0.27	0.17	0.61	0.63
1-35 d	0.97	0.79	1.27	0.49	0.34	0.28	0.80	0.70

[1] Pooled standard error of the mean.
[2] monoglycerides 0.05% in diet
[3]salinomycin 0.006% in diet
[4]Interaction of monoglycerides and salinomycin.

Table 6. Effect of monoglyceride with salinomycin interaction on intestine morphology in broiler chickens at 35 days

Ithems	MGB		SLM		SEM[1]	*P*-value		
	0.00%	0.05%	0.00%	0.006%		MGB[2]	SLM[3]	MGB × SLM[4]
Duodenum								
Villus height (μm)	1380.77	149.01	1376.33	1498.45	33.05	0.02	0.02	0.01
Crypt depth (μm	290.69	350.23	276.66	362.99	16.10	0.01	<0.01	0.06
VH:CD[5]	4.88	4.23	5.03	4.01	0.20	0.04	<0.01	0.17
Jejunum								
Villus height (μm)	647.03	738.00	686.19	704.59	15.27	<0.01	0.42	0.21
Crypt depth (μm	186.06	227.05	192.82	214.32	8.40	<0.01	<0.01	<0.01
VH:CD	3.67	3.10	3.77	3.06	0.17	<0.01	<0.01	<0.01
Ileum								
Villus height (μm)	705.05	592.54	616.30	658.93	15.07	<0.01	<0.01	1
Crypt depth (μm)	185.60	172.98	178.95	180.47	2.65	0.02	0.75	0.64
VH:CD	3.77	3.47	3.64	3.61	0.08	0.09	0.89	0.56

[1] Pooled standard error of the mean.
[2] monoglycerides 0.05% in diet.
[3]salinomycin 0.006% in diet.
[4]Interaction of monoglycerides and salinomycin.
[5]Villus height to Crypt Depth ratio.

4. Discussion

Nowadays, the poultry and swine industries are focusing on developing the use of medium-chain fatty acids and monoglycerides in both *in vitro* and *in vivo* research to study the properties of action to inhibiting viruses and bacteria that can cause diseases in poultry and swine [4] and there are studies from various research to summarize group of active compounds against pathogens and the results from [7] Regarding the effects of combining monoglycerides with chlortetracycline on the growth performance, intestinal morphology, and gut microbiota of the poultry and found that supplementing 500 gram of chlortetracycline per a ton of feed enhances the growth rate of the poultry of 42 days old statistically significantly (p<0.05) comparing to the control group within age of 22-24 days old. The result of supplementing 500 grams of chlortetracycline per ton of feed with monoglycerides 650 or 1,000 gram per ton of feed is show that growth rate is statistically significantly improved (p<0.05) which in line with the study that shows supplementing monoglycerides 0.05% or salinomycin 0.06% in poultry's feed in each age period can improve growth rate poultry.

Supplementing monoglycerides with salinomycin has an influence to improve feed conversion ratio within 1-10 days because supplementation of monoglyceride blends in poultry feed has been found to improve intestinal structure and function to have better immunity protecting digestive system from diseases. Intestine also has better performance which be able to effectively absorb nutrients because is the property of salinomycin that effective prevents some Gram-positive bacteria and inhibits *Clostridium perfringens* that can cause poultry's intestine to inflame which is called necrotic enteritis or NE. However, besides having a better growth rate or higher weight, environment is also a factor to improve performance of monoglycerides [11]

The absorption of monoglycerides occurs directly in the mid and distal parts of the small intestine, where they are then broken down in the mucosa cells in the intestine to butyric acid and be absorbed into blood stream (Because lymphatic system is not developed well in poultry. And then they are metabolized in the liver. [8] For the rest of monoglycerides within distal part of intestine will be converted to butyric acid by microorganism which will be used as energy for the villi cells of the intestine. Therefore, Monoglycerides help improve the morphology of the intestine and enhance the poultry's immune system [13] which consistent with the results of this experiment: combining monoglycerides with salinomycin together will have influence on Villus heigh in Duodenum area to be higher which will improve absorption's efficiency in this area. However, it is highly possible that monoglycerides had some synergy effect with salinomycin on controlling plus-gram bacteria to decline. However, monoglycerides also have features to prevent viruses and support intestine morphology. Therefore, there may be a synergistic effect to improve FCR and intestinal morphology

5. Conclusion

The supplementation of 0.05% monoglycerides combined with 0.006% salinomycin has a synergistic effect on the feed conversion rate to body weight (days 1-10), the height of villi in the duodenum, and crypt depth, along with the VH:CD ratio in the jejunum, resulting in improved growth performance and morphology of the proximal small intestine in broiler chickens.

References

1. Ahmad, R., Y. H. Yu, K. F. Hua, W. J. Chen, D. Zaborski, A. Dybus, F. S. Hsiao and Y. H. Cheng. Management and control of coccidiosis in poultry A review. Anim Biosci 37(1): 1-15. (2024)
2. Conway., D. P. and M. E. McKenzie. Poultry Coccidiosis Diagnostic and Testing Procedures Third Edition. (2007)
3. Guo, Y. Effects of sodium butyrate on mucosal immunity-associated cells in small intestine of broiler. Chin. J. Anim. Nutr.20:469–474 (2008)
4. Jackman JA, Lavergne TA and Elrod CC Antimicrobial monoglycerides for swine and poultry applications. Front. Anim. Sci. 3:1019320 (2022)
5. Jan, S. L. and G. Shieh A systematic approach to designing statistically powerful heteroscedastic 2 x 2 factorial studies while minimizing financial costs." BMC Med Res Methodol 16(1): 114 (2016)
6. Johansen, C.H.; Bjerrum, L.; Pedersen, K. Impact of salinomycin on the intestinal microflora of broiler chickens. Acta Vet. Scand. 2007, 49, 30.

7. Lin, Q., Liu, Y., Li, L., Huai, M., Wang, Y., Lv, T Dai, Q. Effects of a mixture of mono-glycerides of butyric-, capric-, and caprylic acid with chlortetracycline on the growth performance, intestine morphology, and cecal microflora of broiler birds. Poult Sci, 101(2), 101617 (2022)
8. Muscella, A., E. Stefano, P. Lunetti, L. Capobianco and S. Marsigliante. The Regulation of Fat Metabolism During Aerobic Exercise. Biomolecules 10(12) (2020)
9. Naghizadeh, M., L. Klaver, A. A. Schonherz, S. Rani, T. S. Dalgaard and R. M. Engberg. Impact of Dietary Sodium Butyrate and Salinomycin on Performance and Intestinal Microbiota in a Broiler Gut Leakage Model." Animals (Basel) 12. (2022)
10. Niyomsab Patcharanan , Wiriya Loongyai, Sombat Prasongsook and Theerawit Poeikhampha. The Effect of Mono-Butyrin Supplementation in Diet on Egg Production, Egg Quality, Intestinal Morphology of Laying Hens After Peak Production Period. The 20th Kasetsart University Kamphaeng Sean Campus National Conference, December 2023: 162-174 (2023)
11. Sacakli, P., O. O. Cinar, A. Ceylan, M. S. Ramay, J. A. Harijaona, A. G. Bayraktaroglu, Y. Shastak and A. Calik Performance and gut health status of broilers fed diets supplemented with two graded levels of a monoglyceride blend. Poult Sci 102(2): 102359 (2023)
12. Sampugna, J., J. G. Quinn, R. E. Pitas, D. L. Carpenter, and R. G. Jensen. Digestion of butyrate glycerides by pancreatic lipase. Lipids 2:397–402 (1967)
13. Sikandar, A., H. Zaneb, M. Younus, F. Khattak, S. Masood, A. Aslam, S. Ashraf, H. Rehman and M. S. Yousaf Effect of sodium butyrate on performance, immune status, microarchitecture of small intestinal mucosa and lymphoid organs in broiler chickens." Asian-Australasian Journal of Animal Sciences 30 (2017)
14. Srinual, O., M. Punyatong, T. Moonmanee, P. Intawicha, M. Yachai and W. Tapingkae Replacement of Fish Meal with Suckermouth Armored Catfish and Its Effect on Performance and Intestinal Morphology of Indigenous Thai Chicken. Journal of Animal and Plant Sciences 30: 803-810 (2020)
15. Watt, M. J., and G. R. Steinberg. Regulation and function of triacylglycerol lipases in cellular metabolism. Biochem. J.414:313–325 (2008)

Effect of *Palisada perforate* (Bory) K.W.Nam supplementation on gas production and fermentation kinetics in total mixed ration

Nur Hidayah [1,2], Kustantinah[3], *Cuk Tri* Noviandi[3], and *Andriyani* Astuti[3*]

[1]Graduate student of Animal Science. Universitas Gadjah Mada. Jl. Agro Karangmalang. Yogjakarta 55281. Indonesia

[2]Department of Animal Science. Agriculture Faculty. Tidar University. Jl. Barito 1. Magelang. Central Java 56114. Indonesia

[3]Animal Science Faculty. Universitas Gadjah Mada. Jl. Agro Karangmalang. Yogjakarta 55281. Indonesia

Abstract. This research aimed to evaluate *Palisada perforate* (Bory) K.W.Nam as a supplement on total mixed ration to their gas production and fermentation kinetics. A factorial randomized block design with two factors and five replications. The first factor was kinds of total mixed ration (TMR 1, 80% grass (G) and 20% concentrate (C) based on dry matter (DM); TRM 2, 70% G and 30% C; TMR 3, 60% G and 40% C) and the second factor was level of *P. perforate* (Bory) K.W.Nam supplementation (0 and 5% organic matter (OM)). There was no interaction between kinds of TMR and level of *P. perforate* (Bory) K.W.Nam supplementation on gas production and fermentation kinetics. The TMR 3 increased ($P<0.01$) gas production in early incubation (2 and 12 h) but was similar to all treatments on 24 and 48 h of incubation. Meanwhile, increased concentrate proportion on TMR 2 and 3 increased ($P<0.01$) easily degraded fraction (a). Supplementation of *P. perforate* (Bory) K.W.Nam up to 5% OM on all TMR doesn't affect gas production and fermentation kinetics. TMR consisted of 60% grass and 40% concentrate is the best basal diet to increase gas production in early incubation and easily degraded fraction on fermentation kinetics.

1. Introduction

In general, seaweed or macroalgae are aquatic organisms classified with specific characteristics of not having true roots, stems, and leaves [1]. Darmawati et al. [2] explained that in the plant kingdom, seaweed is included in the *Thallophyta* division with all parts called thallus. While the part that resembles a root is called a holdfast, the stipe is the stem, and the blade is the leaf part [3]. Tanna and Mishra [4] stated that seaweeds are incredibly diverse in terms of species, habitat, maturity, harvesting period, and environmental factors, and they are also great sources of various nutrients.

Seaweeds are divided into three groups: red seaweed (Rhodophyta), brown seaweed (Ochrophyta, Phaeophyceae), and green seaweed (Chlorophyta). This classification is based

* Corresponding author: andriyaniastuti@ugm.ac.id

on several characteristics, including pigmentation, other morphological features, photosynthetic membrane structure and elements, and the chemical nature of photosynthetic storage products. Seaweeds get their color from their chlorophyll content; green seaweed color from chlorophyll a and b which amounts to higher plants. Phycoerythrin and phycocyanin are dominant in red seaweed. Whereas, xanthophyll and fucoxanthin pigments, which responsible for the color of brown seaweed [5]. Tanna and Mishra [6] also explained that seaweed is a rich source of bioactive compounds, including dietary fibers, proteins, minerals, and vitamins. Additionally, they contain bioactive substances that may have therapeutic applications in treating inflammation, cancer, oxidative stress, allergies, diabetes, thrombosis, obesity, hypertension, lipidemia, and many degenerative diseases. These substances include polysaccharides, polyphenols, and phytochemicals.

In the animal sector, using seaweed as a feed supplement and substitution increases growth rate, feed conversion efficiency, and ruminant productivity [7]. Besides that, it has been proven very effective and efficient in decreasing ruminant methane emissions and increasing productivity [8, 9, 10]. Bromoform, one of the volatile halogenated compounds found in seaweed, has contributed to this effect. Allen et al. [11] and Costa and Leigh [12] explained that bromoform inhibits or blocks the activity of key methanogenic enzymes (methyl coenzyme M reductase (MCR) and coenzyme M transferase). The halogenated molecules have different molecular weights and are kept in gland cells. These cells release the molecules into the environment to act as a chemical defense against bacteria that are antagonistic to herbivores [13]. Various biological activities, including antioxidant, anti-inflammatory, antiprotozoal, and antibacterial effects, have been demonstrated for these compounds [14]. Ahmed & Nishida [15] stated the anti-methanogenic efficaciousness may also be attributed to additional secondary metabolites. Seaweed also produces a wide range of other non-volatile metabolites, including flavonoids, halogen compounds, and other phenolic compounds [14].

In Indonesia, studies on using various tropical seaweed species for ruminants are still very limited, although it has the highest diversity and is the second producer of seaweed in the world [16, 17]. This is due to a number of factors, including over-reliance on a small number of species, difficulties with sustainable farming and harvesting, and the requirement for technological advancements to exploit a wider variety of species [18, 19]. The two most often utilized seaweed species are *Gracilaria* spp. and *Euchema* spp. Thus, a large number of seaweed species remain unexplored [1]. Kasanah et al. [1] also explained that one of the coasts in Java's South Coast region is the Gunungkidul Yogyakarta area, Indonesia. Its 13 km of beach has the potential to be diverse and has an abundance of natural seaweed sources. One of them is red seaweed, *Palisada perforate* (Bory) K.W.Nam, which has the potential as a supplement or feed additive to reduce ruminant methane production [20]. It should be noted that the result is based solely on seaweed as a single ingredient and not on its addition to basal feed. Because of that, the purpose of this study is to assess the impact of *Palisada perforate* (Bory) K.W.Nam supplementation on total mixed rations (TMR) as basal diets on gas production and fermentation kinetics. This variables are important to evaluate as an indicator for determining rumen fermentation patterns and feed degradation process by rumen microbes.

2. Materials and Methods

2.1. Experimental design

This study used a factorial randomized block design with two factors and five replications, which a time of rumen fluid collection as a block. The first factor was the kinds of total mixed ration (TMR 1, 80% Napier grass (G) and 20% concentrate (C) based on dry matter (DM); TRM 2, 70% G and 30% C; TMR 3, 60% G and 40% C) and the second factor was level of *P. perforate* (Bory) K.W.N supplementation (0 and 5% OM).

2.2. Preparing *Palisada perforate* (Bory) K.W.Nam and total mixed ration

Natural red seaweed, *P. perforate* (Bory) K.W.Nam, was harvested in October 2022 from Gunungkidul, Yogyakarta, Indonesia. To get rid of grit, sand, and other seaweed species, the seaweed was thoroughly cleaned with water. Afterward, the sample was dried on a freeze dryer (Buchi, Lyovapor, L-200) and grounded with a Willey mill until a fine powder (80–100 mesh).

The TMR consisted of grass (*Pennisetum purpureum* cv. Gama Umami) and concentrate (rice brand, wheat pollard, soybean meal, coconut cake meal, corn gluten feed, molasses, and premix). The sample dried in an oven at 55°C for 3 days and grounded until powder formed with a Willey mill (80–100 mesh). The concentrate formula and nutrient content of TMR are shown in Tables 1 and 2. The nutrient content of TMR as recommended by Kearl [21] as follows crude protein 10.69% and total digestible nutrient 69.93% based on DM. The TMR nutrient content has iso crude protein and iso total digestible nutrient.

Table 1. Concentrate formula

Ingredients (%)	**Concentrate 1**	**Concentrate 2**	**Concentrate 3**
Rice bran	13.25	17.25	20.25
Wheat pollard	45	45	45
Soybean meal	11.25	7.25	4.25
Coconut cake meal	14.50	14.50	14.50
Corn gluten feed	10	10	10
Molasses	5	5	5
Premix	1	1	1
Total	100	100	100

Table 2. TMR nutrient content (%)

TMR / Nutrient	1 (80% G + 20% C)	2 (70% G + 30% C)	3 (60% G + 40% C)
OM	83.85	83.60	86.48
Ash	15.95	16.10	13.52
CP	13.03	13.03	12.91
EE	2.64	3.13	2.62
CF	28.88	26.45	27.52
NFE	41.02	43.52	43.43
TDN	58.79	61.35	63.23

TMR: total mixed ration; G: grass; C: concentrate; OM: Organic matter; CP :Crude protein ; EE : Ether extract ; CF : Crude fiber ; NFE : Nitrogen free extract ; TDN : Total digestible nutrient ; NFE = 100 – (% CP + % CF + % EE + % ash) ; TDN = 70.6 + 0.259 CP + 1.01 EE - 0,76 CF + 0,091 NFE (Sutardi, 2001)

2.3. *In vitro* process and gas production evaluation

In vitro process, we used a piston and glass syringe of 100 mL according to the Menke & Steingass [22] method. The glass syringe was filled with rumen liquor and buffer solution (total at 30 mL, 1:2 ratio vol/vol), and also 200 mg sampled (TMR with or without *P. perforate* (Bory) K.W.Nam supplemented) as substrates. Using a vacuum tube, the rumen liquor was collected and mixed from the Bali cow and bull fistulated. The Animal Care Committee, Veterinary Faculty, Universitas Gadjah Mada, Indonesia approved the rumen collection process. The *in vitro* process was carried out under anaerobic conditions until 48 h of incubation. We evaluated and calculated total and kinetic gas production with the Neway program [23].

2.4. Statistical analysis

ANOVA was used for statistical analysis, and Duncan's multiple range test (DMRT) was used to look at treatment mean differences. The statistical analysis was performed using IBM SPSS Statistics version 26.

3. Result and Discussion

The result showed that there was no interaction between kinds of TMR and level of *P. perforate* (Bory) K.W.Nam supplementation on gas production and fermentation kinetics. The TMR 3 increased ($P<0.01$) gas production in early incubation (2 and 12 h) but was similar to all treatments on 24 and 48 h of incubation (Table 3). Meanwhile, increased concentrate percentage on TMR 2 and 3 increased ($P<0.01$) easily degraded fraction (a) (Table 4). Supplementation of *P. perforate* (Bory) K.W.Nam up to 5% OM on all kinds of TMR doesn't affect gas production and fermentation kinetics.

Increasing the proportion of concentrate and decreasing the grass proportion in the TMR as a basal diet significantly increased the gas production in early incubation (2 and 12 h). This result indicated that increasing concentrate enhances rumen activities, gas production, and improves feed degradability in early incubation. This result is linear with the easily degraded fraction on fermentation kinetics variables, which increased with increasing concentrate proportion. This described that increasing concentrate followed the easily degradable component that gives more substance for rumen microbial to degrade.

The concentrate has a more easily degradable component in the rumen than grass. The crude protein (CP) and nitrogen-free extract (NFE) of concentrate were higher than grass, 8.98% vs 18.90% and 39.36% vs 53.94% DM, respectively. Opposite to the rumen's difficult-degradable component, grass is higher than concentrate. Grass has crude fiber (CF) content higher than concentrate, 33.42% vs 18.67% DM. Kulivand & Kafilzadeh [24] stated that CP (easily degradable component) has a positive correlation ($r = 0.62$, $p < 0.05$) with gas production at 96 h incubation. Whereas, a negative correlation is observed with gas production when the difficult degradable component, like acid detergent fibre (ADF) that crude fiber part, ($r = -0.60$, $p < 0.05$). Similar results reported by Viennasay et al. [25], increased concentrate levels on a basal diet that consisted of roughage and concentrate significantly increased rumen gas production.

Supplementation of *P. perforate* (Bory) K.W.Nam up to 5% OM on TMR based as basal diet did not increase gas production and fermentation kinetics. This indicated that the percentage of seaweed did not alter microbial activity and feed degraded in the rumen. The same result was reported by Chot et al. [26], after 12, 24, and 48 h of incubation, seaweed extract supplementation reduced methane yield but did not change significantly overall gas production. The result of this research also described that the seaweed species and amount of supplementation did not disturb the microbial activity to feed degraded. A study by Künzel et al. [27] reported that the inclusion of 2.5 and 5.0% on a DM basis brown seaweed from the Icelandic ocean (*Ascophyllum nodosum* and *Fucus vesiculosus*) on total mixed ration (TMR) reduced total gas production with 5% level ($P < 0.001$). The different response might be due to the kind and percentage of secondary metabolite from the seaweed. In this research, we used red seaweed.

Table 3. Gas production with different levels of *Palisada perforate* (Bory) K.W.Nam supplementation on different kinds of total mixed ration

Variable / Time incubations		Gas production (mL/200 mg DM)			
		2 h	12 h	24 h	48 h
Interaction kinds of total mixed ration (TMR) with different levels of *P. perforate* (Bory) K.W.Nam supplementation					
TMR 1	0% OM	7.75 ± 1.89	26.35 ± 4.98	42.23 ± 5.26	53.50 ± 7.16
	5% OM	6.82 ± 1.95	25.77 ± 3.38	41.03 ± 4.60	50.54 ± 7.26
TMR 2	0% OM	6.23 ± 2.64	26.71 ± 5.89	38.29 ± 7.56	47.48 ± 7.35
	5% OM	9.03 ± 1.69	28.26 ± 2.81	39.98 ± 5.74	51.63 ± 5.53
TMR 3	0% OM	10.27 ± 2.23	31.17 ± 3.14	45.82 ± 3.04	57.84 ± 2.37
	5% OM	9.22 ± 2.28	29.22 ± 3.63	42.82 ± 4.12	54.13 ± 4.93
Effect of different kinds of TMR					
TMR 1		7.34[a] ± 1.86	26.09[a] ± 4.09	41.70 ± 4.71	51.86 ± 6.93
TMR 2		7.63[a] ± 2.54	27.40[a] ± 4.58	39.14 ± 6.39	49.79 ± 6.35
TMR 3		9.68[b] ± 2.18	30.09[b] ± 3.36	44.15 ± 3.80	55.76 ± 4.25
Effect of different levels of *P. perforate* (Bory) K.W.Nam supplementation					
0% OM		8.06 ± 2.64	27.86 ± 5.03	41.85 ± 6.16	52.94 ± 7.07
5% OM		8.42 ± 2.15	27.86 ± 3.39	41.30 ± 4.67	52.10 ± 5.76

Different superscripts in the same rows show significant differences ($P<0.05$); TMR 1: 80% G + 20% C; TMR 2: 70% G + 30% C; TMR 3: 60% G + 40% C; G: grass; C: concentrate; OM: organic matter

Table 4. Fermentation kinetics with different levels of *Palisada perforate* (Bory) K.W.Nam supplementation on different kinds of total mixed ration

Variable		Kinetic gas production			
		a (mL/200 mg DM)	b (mL/200 mg DM)	c (mL/jam)	a+b (mL/200 mg DM)
Interaction kinds of total mixed ration (TMR) with different levels of *P. perforate* (Bory) K.W.Nam supplementation					
TMR 1	0% OM	1.87 ± 2.90	57.30 ± 7.31	0.055 ± 0.01	59.17 ± 9.93
	5% OM	0.34 ± 2.78	58.50 ± 5.85	0.043 ± 0.01	58.84 ± 7.85
TMR 2	0% OM	4.66 ± 4.19	57.45 ± 10.11	0.042 ± 0.00	62.10 ± 7.45
	5% OM	4.27 ± 1.10	56.37 ± 3.80	0.045 ± 0.00	60.64 ± 4.83
TMR 3	0% OM	5.05 ± 1.84	60.15 ± 1.00	0.047 ± 0.00	65.19 ± 1.46
	5% OM	4.17 ± 2.23	59.71 ± 1.40	0.046 ± 0.00	63.88 ± 2.23
Effect of different kinds of TMR					
TMR 1		1.02[a] ± 2.77	57.97 ± 6.13	0.048 ± 0.01	58.99 ± 8.23
TMR 2		4.44[b] ± 2.69	56.85 ± 6.77	0.043 ± 0.00	61.29 ± 5.75
TMR 3		4.61[b] ± 1.98	59.93 ± 1.17	0.047 ± 0.00	64.54 ± 1.90
Effect of different levels of T*P. perforate* (Bory) K.W.Nam supplementation					
0% OM		3.95 ± 3.12	58.44 ± 6.42	0.048 ± 0.01	62.39 ± 6.78
5% OM		2.93 ± 2.75	58.19 ± 4.06	0.045 ± 0.00	61.12 ± 5.51

Different superscripts in the same rows show significant differences ($P<0.05$); TMR 1: 80% G + 20% C; TMR 2: 70% G + 30% C; TMR 3: 60% G + 40% CC; G: grass; C: concentrate; OM: organic matter

4. Conclusion

It is concluded that TMR consisted 60% grass and 40% concentrate is the best basal diet to increase total gas in early incubation and easily degraded fraction on fermentation kinetics. Using this basal diet has the potential to reduce methane production because it is more degradable in the rumen. This is good for increasing ruminant productivity and reducing environmental pollution. Future research needs to examine the higher level of this seaweed supplementation or another seaweed species on this basal diet.

5. Acknowledgments

We would like to express our gratitude and appreciation to Indonesia Endowment Funds for Education (LPDP) Center for Higher Education Funding (BPPT) for supporting this research.

6. References

1. N. Kasanah, D.S. Seto, H. Khotimah, and Triyanto, Rumput laut pangan (Universitas Gadjah Mada Press, Yogyakarta, 2022)
2. Darmawati, I. Sutinah, Ode, B.D.H. Setyono, S. Laheng, T. Mujtahidah, Y.P. Sari, I. Wahyuni, Z. Abidin, W. Sukendar, S.T.M. Sulthoniyah, A.B. Marda, M.A. Yusuf, A.N.N. Suci, A. Abdullah, and V. Setyaka, Kiat agribisnis rumput laut (CV. Tohar Media, Makasar, 2023)
3. C.M. Buchholtz, G. Krause, and B.H. Buch, Seaweeds and man in seaweed biology ecological studies (Springer-Verlag, Berlin, 2012)
4. B. Tanna, A. Mishra, Compr. Rev. Food Sci. Food Saf. **17**, 1613–1624 (2018)
5. B. Kılınç, S. Cirik, G. Turan, H. Tekogul, E. Koru, Seaweeds for food and industrial applications, *In Food industry*, ed. I. Muzzalupo (Intech Open, London, 2013)
6. B. Tanna, A. Mishra, Compr. Rev. Food Sci. Food Saf. **18**, 817–831 (2019)
7. B.R. Min, D. Parker, D. Brauer, H. Waldrip, G. Lockard, K. Hales, A. Akbay, S. Augyte, Anim. Nutr. **7**, 1371-1387 (2021)
8. B.M. Roque, C.G.J. Brooke, J. Ladau, T. Polley, L.J. Marsh, N. Najafi, P. Pandey, L. Singh, R. Kinley, J.K. Salwen, E. Eloe-Fadrosh, E. Kebreab, and M. Hess. Anim. Microbiome **1**, 1-3 (2019)
9. R.D. Kinley, G. Martinez-Fernandez, M.K. Matthews, R. de Nys, M. Magnusson, and N.W. Tomkins. J. Clean Prod. **259**, 120836 (2020)
10. A.A. Mihaila, C.R.K. Glasson, R. Lawton, S. Muetzel, G. Molano, and M. Magnusson. Appl. Phycol. **3**, 274-284 (2022)
11. K.D. Allen, G. Wegener, R.H. White, Appl. Environ. Microbiol. **80**, 6403-6412 (2014)
12. K.C. Costa, J.A. Leigh, Curr. Opin. Biotechnol. **29**, 70-75 (2014)
13. C. Parchemin, D. Raviglione, H. Ghosson, M.V. Salvia, C. Goossens, P. Sasal, E. Faliex, N. Tapissier-Bontemps, Algal Res. 72, 103138 (2023)
14. F.S.-L. Keng, S.-M. Phang, N. Abd Rahman, E.C. Leedham Elvidge, G. Malin, W.T. Sturges, J. Appl. Phycol. **32**, 1377-1394 (2020)
15. E. Ahmed, T. Nishida, Anim. Feed Sci. Technol. **318**, 116118 (2024)
16. Food and Agriculture Organization, The global status of seaweed production, trade and utilization (Food and Agriculture Organization of the United Nations, Rome, 2018)
17. E. Erniati, F.R. Zakaria, E. Prangdimurti and D.R. Adawiyah. Acta Aquatica. **3**, 12-17 (2016)
18. M. MacMonagail. L. Cornish. L. Morrison. R. Araújo. and A.T. Critchley, Eur. J. Phycol. **52**. 371-390 (2017)
19. S. Kraan. Seaweed resources. collection. and cultivation with respect to sustainability. In *Sustainable Seaweed Technologies* (Elsevier Inc., United Kingdom, 2020).

20. N. Hidayah, C.T. Noviandi, A. Astuti, and Kustantinah. Nusantara Biosci. **16**, 37-42 (2024)
21. L.C. Kearl, Nutrient requirements of ruminants in developing countries (International Feed Stuffs Institute, Utah Agriculture Experimental Station, Utah State University, Utah, 1982)
22. H.H. Menke, H. Steingass, Anim. Res. and Dev. **28**, 7-55 (1988)
23. X.B. Chen, Neway program international feed resources unit (Rowett Research Institute, Backburn, 1994)
24. M. Kulivand, F. Kafilzadeh, Acta Sci. **31**, 9-14 (2015)
25. B. Viennasay, P. Totakul, M. Matra, B. Phestcha, M. Wanapat, Agric. Food Sci. **102**, 4927-4932 (2022)
26. Y. Choi, S.J. Lee, H.S. Kim, J.S. Eom, S.U. Jo, L.L. Guan, . . .S.S. Lee, Sci. Reports. **11**, 24092 (2021)
27. S. Künzel, T. Yergaliyev, K.J. Wild, H. Philippi, A.H. Petursdottir, H. Gunnlaugsdottir, . .M. Rodehutscord, Front. Microbiol. **13**, 889618 (2022)

Nutrient content of jack bean (Canavalia ensiformis) at different growth stages in Blora, East Java, Indonesia

Bayu Prasetyo[1], *Bambang* Suwignyo[1], *Genki* Ishigaki[2], *Takahiro* Gondo[3], *Diafan* Kurnia Jati[4] and *Yogi* Sidik Prasojo[1*]

[1]Department of Animal Nutrition and Feed Science, Faculty of Animal Science, Universitas Gadjah Mada, Yogyakarta, 55281, Indonesia
[2]Sumiyoshi Livestock Science
[3]Frontier Science Research Center, University of Miyazaki, 1-1 Gakuen Kibanadai-nishi,
[4]Miyazaki,889-2192,Japan
Pertamina Foundation, Pertamina Learning Center, South Jakarta, Jakarta 12220, Indonesia

Abstract. This study aims to determine the effect of harvesting at different growth stages on nutrient content of jack bean plants cultivated in alluvial type of soil at Blora, Central Java, Indonesia. A field experiment was conducted using a completely randomized block design with four replications. Plant in every plot were randomly selected and hand separated. Each component was sampled approximately 1000 g. Data were analyzed using Analysis of Variance (ANOVA) and continued with Duncan's Multiple Range Test ($p<0.05$). The following parameter observed were nutrient content include dry matter (DM), organic matter (OM), crude protein (CP), crude fiber (CF), dan ether extract (EE). The result showed that growth stages had a significant effect ($P<0,05$) on DM and CF, very significantly ($P<0,01$) on OM and CP, while the CF and EE not significantly ($P>0,05$). DM and CF content increased during the flowering growth stage. On the other hand, the OM content increased as the growth stage increased. However, the CP and EE content decreased as growth stages increased. The vegetative stage is considered the best harvesting stage because it has the highest crude protein content, as leguminous plants are forage plants and an important source of protein for ruminant feed.

1 Introduction

Ruminant livestock are a type of livestock with the unique characteristic of having a stomach consisting of four compartments: the reticulum, rumen, omasum, and abomasum. This uniqueness enables ruminants to utilize forage-based feed to meet their basic maintenance, production, and reproduction needs. Generally, ruminant livestock are fed forage in the form of grasses, which have low protein content [1], necessitating the use of forage with higher protein content. One type of forage that can serve as a protein source for ruminants is legumes.

* Corresponding author: yogi.sidik.p@ugm.ac.id

Indonesia is a country rich in local legume plant resources, have great potential, and many of which remain under-researched. One potential legume plant is jack bean (*Canavalia ensiformis*). Jack bean is a crop distributed across tropical regions and has demonstrated good adaptability, including in Indonesia. Jack bean is a diversified crop with several advantages, including drought tolerance, resistance to pests, high productivity, and versatility for various uses such as direct food consumption, raw material for the food industry, pharmaceutical industry, livestock feed, and as green manure [2]. The productivity of jack bean seeds is relatively high compared to soybeans, with yields of 8–9 tons/ha and 1.2–2 tons/ha, respectively [2]. Jack bean seeds contain nutrients including moisture 7.57%, protein 32.02%, fat 3.83%, ash 4.39%, crude fiber 5.01%, and carbohydrates 46.29%, with seeds sourced from the Genetic Resources Unit of the International Institute of Tropical Agriculture (IITA), Ibadan, Nigeria [3].

Jack bean plants can grow up to 2000 meters above sea level in tropical areas with temperatures ranging from 20 to 32°C. These plants can thrive in dry areas due to their deep root systems, which allow them to absorb more water compared to grasses. In addition to their seeds, jack bean plants offer other potential uses, particularly for ruminant feed, through their biomass. The biomass yield of Canavalia ensiformis is approximately 21.81 tons/ha fresh weight and 11.29 tons/ha dry matter under tropical environmental conditions, with annual rainfall of 1200 mm and temperatures between 19–33°C [4].

Legume plants generally undergo three growth stages: vegetative, flowering, and pod-setting. The vegetative stage is the period of intensive growth involving vegetative organs such as leaves, stems, and roots. The flowering stage, or the emergence of flowers, marks the transition from the vegetative stage to the opening of the first flower. The pod-setting stage is the period from the end of flowering to the filling of pods on the oldest flower cluster [5]. Harvesting at different growth stages of legume plants impacts their morphological characteristics and nutrient composition [6].

Research on the utilization of jack bean (*Canavalia ensiformis*) as feed for ruminants is still limited both in Indonesia and internationally. The productivity and quality of jack bean plants are influenced by several factors, such as environmental conditions and the growth stage at the time of harvest. Harvesting at different growth stages of jack bean plants could be a breakthrough for the livestock industry in utilizing jack bean as a forage feed source of protein beneficial for ruminants. Therefore, this study aims to determine the effect of harvesting at each growth stage on the dry matter production and nutrient content of jack bean plants.

2 Materials and methods

2.1 Plant materials and experimental site

Jack bean (Canavalia ensiformis) Setren accession seeds were used as plant material for this study. The study was carried out at an elevation of 100 meters above sea level in Megeri Village, Kradenan District, Blora Regency, Central Java (7º21'40''S 111º26'36''E) (Figure 1). Most of the land in Blora Regency is dry land. The soil type characterized as alluvial. The climate of Blora Regency is tropical climate, relatively high temperatures in the dry season and sufficient rainfall in the rainy season. Precipitation and air temperature data for the location in 2023 (Figure 2) were obtained from the Ministry of Public Works and People's Housing (PUPR) Bengawan Solo Data for Ngawi Regency, East Java (URL: Sistem Informasi Hidrologi & Kualitas Air (bbws-bsolo.net)).

Table 1. Soil characteristics and physical properties in the experimental site

Soil Parameters[1]	Value
pH (H_2O)	7,59
C Organic (%)	0,50
N Total (%)	0,12
Available P (ppm)	1,33
Available K (Me%)	0,72

[1]C: carbon; N: nitrogen; P: phosphorus; K: potassium

2.2 Climatic condition

The precipitation and average air temperature during the research in 2023 can be seen in Table 1. The lowest rainfall during the crop maintenance period occurred in September and the highest in November, while the lowest and highest air temperatures occurred in September and October, respectively.

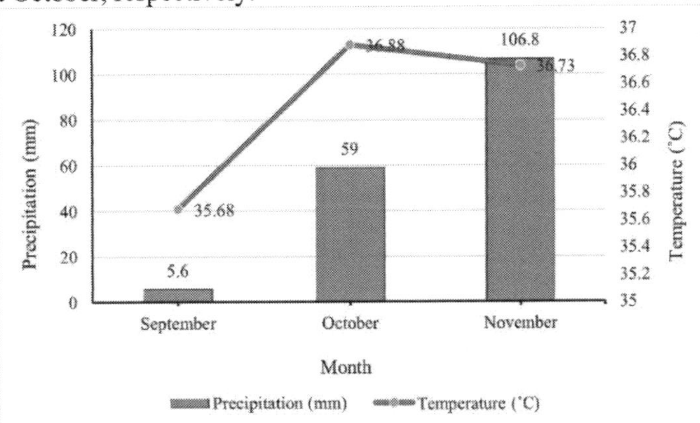

Fig. 1. Climate conditions in term of precipitation and temperature at the experimental site

2.3 Experimental design and cultivation

Planting was carried out for 55 day after planting (DAP) (vegetative), 70 DAP (flowering), and 90 DAP (pod-setting) from September to November 2023. The research area employed a randomized block design with three treatments and four replications. Each replication consisted of plots measuring 1 m^2 (1x1 m) with a total of 12 plots and a spacing of 0.5 m between plots, comprising three growth stages: vegetative, flowering, and pod development. Planting was done by making holes in the soil approximately 2-3 cm deep, with one jack bean seed planted per hole at a spacing of 25x25 cm, in 4 rows, resulting in a plot density of 16 seeds/m^2. Plant maintenance included watering and weeding, with watering conducted once a day at 05:00 AM if there was no rainfall, and weeding done weekly.

2.4 Measurement of cutting and yield

The plant sample cutting was performed 14 cm from the soil surface in different plant growth stages, vegetative (55 DAP), flowering (70 DAP), and pod-setting (90 DAP). The cut plants were then weighed using a hanging scale to obtain fresh weight (leaves and stems). Each component was sampled approximately 1000 g and put into a paper bag and then oven-dried at 55°C for 4 days and weighed back to obtain the dry weight. Dried samples were ground to pass a 1.0 mm screen using a Wiley mill and then stored individually inside the sealed plastic bag based on the treatment. Ground samples were used for further laboratory analysis.

2.5 Laboratory analysis

The chemical analysis of nutrient content included dry matter (DM), organic matter (OM), crude protein (CP), crude fiber (CF), and ether extract (EE) was conducted based on [7].

2.6 Statistical analysis

Statistical analysis was used to compare the results of morphological characteristics and biomass production among different growth stages. The data were statistically analysed using analysis of variance (ANOVA) followed by Duncan's Multiple Range Test (DMRT) if significant differences were found (at a significance level of 95%) using SPSS 23.0 (SPSS Inc., USA).

3 Result and Discussion

The analysis results of the nutrient content of jack bean (*Canavalia ensiformis*) plants at each growth stage include dry matter, organic matter, crude protein, crude fiber, and ether extract. The data on the differences in nutrient content of jack bean plants at different growth stages can be seen in Table 2.

Table 2. Nutrient content in jack bean (*Canavalia ensiformis*) at different growth stage (Means±SE)

Nutrient (%)	Growth Stage		
	Vegetative	Flowering	Pod-setting
Dry matter	$20{,}941^{a}\pm0{,}312$	$22{,}098^{b}\pm0{,}419$	$20{,}168^{a}\pm0{,}320$
Organic matter*	$83{,}577^{a}\pm0{,}323$	$85{,}813^{b}\pm0{,}763$	$87{,}350^{b}\pm0{,}348$
Crude protein*	$21{,}656^{b}\pm0{,}746$	$17{,}276^{a}\pm0{,}524$	$17{,}320^{a}\pm0{,}686$
Crude fiber	$19{,}910^{a}\pm0{,}407$	$21{,}946^{b}\pm0{,}599$	$19{,}392^{a}\pm0{,}717$
Ether extract^{ns}	3,675±0,665	3,204±0,962	2,818±0,910
Total digestible nutrient*	$54{,}147^{a}\pm0{,}931$	$53{,}490^{a}\pm0{,}722$	$60{,}593^{b}\pm1{,}512$

ns : non significant
ab : Different superscripts on the same line indicate significant differences (P<0,05)
* : Different superscripts on the same line indicate highly significant differences (P<0,01)

The analysis of the total dry matter content of jack bean (*Canavalia ensiformis*) plants showed different results between leaves and stems. The total dry matter content increased during the flowering growth stage (22.098%) compared to the vegetative stage (20.941%), and then significantly decreased (P<0.05) during the pod-setting stage (20.168%). In the early reproductive stage, such as flowering, the production of leaves and stems reaches its peak before entering the pod-setting stage. [8] stated that in legumes during the early reproductive stage, such as flowering, the photosynthetic products generated by the leaves are distributed and utilized for the formation of primary leaves and stems, thereby increasing the plant's dry matter content. [6] stated that harvesting legume plants, such as soybeans, should be conducted before entering the R7 stage, when one of the pods on the main stem begins to turn brown. Once the plant has entered this stage, the number of leaves and branches decreases, leading to a reduction in dry matter content. Furthermore, the stems become hard and woody, reducing the quality of legumes used as forage crops.

The analysis of total organic matter content showed a highly significant increase ($P<0.01$) with the advancement of growth stages in jack bean (*Canavalia ensiformis*) plants. This is attributed to the decomposition of plant parts that detach, thereby enhancing soil nutrient content, such as soil organic carbon. [9] stated that soil organic carbon content is crucial for plants, especially for their nutrient content. Soil organic carbon, such as humus and other organic compounds, serves as the primary energy source for soil microbes. Soil microbes play an important role in the decomposition process of organic matter, which produces essential nutrients like nitrogen and phosphorus. [10] noted that one example of soil microbes involved in the decomposition of organic matter is arbuscular mycorrhizal fungi (AM). These fungi play a role in accelerating the decomposition process of organic matter through enzymatic activity and releasing nutrients from organic matter, such as inorganic nitrogen and phosphorus, which can then be utilized by plants. These nutrients become available and can be absorbed by plants for the formation of plant organic matter through the process of photosynthesis.

The analysis results showed that the crude protein content in leaves and total crude protein significantly decreased ($P<0.01$) as the growth stage of jack bean (*Canavalia ensiformis*) plants progressed. The highest results were obtained during the vegetative stage, with a decrease observed in the flowering and pod-setting stages. These findings align with studies on soybean plants [6], sunn hemp (*Crotalaria juncea*) [11], alfalfa [12], and pigeon pea (*Cajanus cajan*) [13], which showed that crude protein content in leaves and total decreased with advancing growth stages or later cutting ages. This decline can be attributed to the fact that, during the vegetative stage, jack bean plants store a significant amount of crude protein from photosynthesis, which is distributed to plant organs, particularly the leaves. The high protein content in the leaves during the vegetative stage contributes to the total crude protein content in the same stage. These results are consistent with findings by [14], who reported that alfalfa plants in the vegetative growth stage exhibited the highest crude protein content, which decreased with advancing growth stages. During the vegetative stage, legume plants produce more protein through photosynthesis, resulting in higher protein content. [15] stated that, during the reproductive stage to maturity, protein content decreases due to a reduction in leaf and stem production. As the flowering stage begins, nitrogen stored during the vegetative stage is redistributed to other parts of the plant for the formation of flowers and pods, leading to a decrease in the proportion of crude protein (NPN) in the plant.

The analysis results of the total fiber content of jack bean (*Canavalia ensiformis*) plants showed significant differences ($P>0.05$) at each growth stage. The highest results were obtained at the flowering stage compared to the vegetative and pod-setting stages. The increase in crude fiber content during the flowering stage was due to the proportion of structural tissue in the stem and leaves. The proportion of structural tissue is greater in the stem than in the leaves, which are mostly occupied by thin-walled mesophyll cells during plant maturation. High fiber content can lead to a decrease in digestibility, resulting in less fermentable energy for ruminant animals. In general, leaves are more digestible because they have fewer cell-wall constituents than stems, so the increase in forage value depends on the proportion of plant parts. The crude fiber content in the stems was higher than in the leaves of jack bean plants. This is due to the high lignification process during plant development, which mainly occurs in the stem, resulting in higher crude fiber content in the stem. Lignin composition is very limited at the early stages of development and begins to mature as the plant ages. Lignin composition increases in the middle lamella's main wall area, which is the least digestible part. The fraction of lignified tissue in legumes increases over time as the stem grows [11].

The results of the total ether extract content analysis showed no significant differences ($P>0.05$) at each growth stage. However, there was a trend of decreasing ether extract values as the growth stage advanced, with the highest results observed during the vegetative stage.

This could be attributed to the leaf-to-stem ratio at each growth stage. [11] stated that leaves of leguminous plants are high in protein and fat content. Therefore, the proportion of leaves greatly influences the ether extract content in plants. [16] reported that ether extract content in plants is determined by the leaf ratio, as the fat and protein in forages mostly originate from the leaves. The ether extract content in plants is correlated with crude protein. The ether extract and crude protein contents in leguminous plants tend to decrease as the growth stage progresses. This aligns with the results of the total crude protein content analysis shown in Table 2.

In conclusion, The dry matter and crude fiber content increased during the flowering growth stage. On the other hand, the organic matter content increased as the growth stage of the plant progressed. However, the crude protein and ether extract content decreased with the advancement of the growth stage. The vegetative growth stage is considered the best harvesting stage because it has the highest crude protein content, as leguminous plants are forage plants and an important source of protein.

References

1. M. R. Islam, S. C. Garcia, N. R. Sarker, Md. A. Islam, and C. E. F. Clark, Napier grass (*Pennisetum purpureum* Schum) management strategies for dairy and meat production in the tropics and subtropics: yield and nutritive value. Frontiers in Plant Science. **14**, 1-15 (2023). https://doi.org/10.3389/fpls.2023.1269976
2. I. Marjuki, S. E. P. Susilaningsih, M. Th. Darini, Pengaruh macam amelioran lokal dan dosis legin koro terhadap pertumbuhan dan hasil tanaman koro pedang (*Canavalia ensiformis* L.) di lahan marjinal tanah grumusol. Jurnal Ilmiah Agroust. **2**, 126-135 (2018).
3. J. O. Odedije, E. A. Akande, O. O. Fapojuwo, and O. J. Olawuyi, Influence of roasting on nutritional and antinutritional factors of jackbean, *Canavalia ensiformis* (L) D.C. flour. Journal of Underutilized Legumes. **1**, 112-121 (2019).
4. P. E. Teodoro, L. P Rubeiro, F. A. da Silva, C. C. G. Correa, R. A. A. da Luz Jr., and F. E. Torres, Sample size to estimate biomass and productivity of *Canavalia ensiformis* and *Dolichos lablab*. *Revista de Ciências Agrárias*. **37**, 348-353 (2014). https://doi.org/10.19084/RCA.16834
5. N. Hidayati, Toleransi daun dan akar *Vicia faba* L. dalam kondisi stress air pada tiga fase pertumbuhan. Jurnal Biologi Indonesia. **2**, 37-45 (1996). https://doi.org/10.47349/jbi
6. Y. S. Prasojo, G. Ishigaki, M. Hashiguchi, and R. Akashi, Effect of different growth stages on biomass weight and frage quality of different growth type soybean (*glycine max*). Asian Journal of Plant Science. **20**, 256-262 (2021). https://doi.org/10.3923/AJPS.2021.256.262
7. AOAC, Official Methods of Analysis. 12th ed. (Association of Official Analytical Chemist, Washington D. C., 2005).
8. R. Bulbaya, and A. W. Lessen, Nutritional Composition of Grain Legume Leaves and the Impact of Leaf Removal on Yield. Agrosystems, Geoscience, and Environment. **2**, 180015 (2019). http://dx.doi.org/10.2134/age2018.06.0015
9. C. Paungfoo-Lonhienne, J. Visser, T. G. A, Lonhienne, and S. Schmidt, Past, present, and future of organic nutrients. Plant Soil. **359**, 1-18 (2012). http://dx.doi.org/10.1007/s11104-012-1357-6

10. A. Hodge and A. H. Fitter, Substantial nitrogen acquisition by arbuscular mycorrhizal fungi from organic material has implications for N cycling. PNAS. **107**, 13754-13759 (2010). https://doi.org/10.1073/pnas.1005874107
11. J. P. Marković, R. T. Štrbanović, D. V. Terzić, D. J. Djokić, A. S. Simić, M. M. Vrvić and S. P. Živković, Changes in lignin structure with maturation of alfalfa leaf and stem in relation to ruminants nutrition. Afr. J. Agric. Res., **7**, 257-264 (2012). https://doi.org/10.5897/AJAR11.1485
12. M. S. Sayar, M. Basbag, E. Cacan, and H. Karan, The effect of different cutting times on forage quality traits of alfalfa (*Medicago sativa* L.) genotypes and evaluations with biplot analysis. Presenius Environmental Bulletin. **31**, 9172-9184 (2022).
13. O. O. Bode, F. A. Noah, and O. O. Jacob, Effects of spacing, cutting height and cutting interval on fodder yield and nutritional value of *Cajanus cajan*. IJEAB. **3**, 818-822 (2018). https://doi.org/10.22161/ijeab%2F3.3.14
14. E. Karayilanli, and V. Ayhan, Investigation of feed value of alfalfa (*Medicago sativa* L.) harvested at different maturity stages. Legume Research. **39**, 237-247 (2016). http://dx.doi.org/10.18805/lr.v0iOF.9292
15. J. Markovic, M. Petrovic, D. Terzic, T. Vasic, I. Kostic, R. Strbanovic, and G. Grubic, Protein fractions as influenced by cultivars, stage of maturity and cutting dates in alfalfa (*Medicago sativa* L.). Legume Research-Am Internasional Journal. **45**, 46-51 (2022). http://dx.doi.org/10.18805/LR-479
16. M. Whetsell, and E. Rayburn, Variation in fatty acids concentration in grasses, legumes, and forbs in Allegheny Plateau. Agronomy. **12**, 1693 (2022). https://doi.org/10.3390/agronomy12071693

Growth of Tropical Alfalfa (*Medicago sativa* cv. Kacang Ratu BW) under Drought Stress

Sri Widiastuti[1]*, *Bambang* Suwignyo[2], and *Asih* Kurniawati[2]

[1]Graduate School of Animal Science, Universitas Gadjah Mada, Jl. Fauna No. 3 Bulaksumur, Yogyakarta, 55281, Indonesia

[2]Departement of Animal Nutrition and Feed Science, Faculty of Animal Science, Universitas Gadjah Mada, Jl. Fauna No. 3 Bulaksumur, Yogyakarta, 55281, Indonesia

Abstract. Tropical alfalfa (*Medicago sativa* cv. Kacang Ratu BW) has potential as a high-quality forage source in tropical regions. This study aimed to observe the response of tropical alfalfa to drought stress by examining plant growth. The experiment employed a Nested Design with two factors: irrigation frequency (P0, P1, P2, and P3) and cutting age (20 and 30 days). Observed variables included plant height, growth rate, and number of branches. Results indicated no significant differences (P>0.05) across all variables at different levels of drought stress and cutting ages. These findings indicate the potential of tropical alfalfa to be cultivated in drought-prone tropical environments. This study provides a scientific basis for developing of more effective management strategies in alfalfa cultivation in the tropics, which may contribute to the increased availability of high-quality feed throughout the year for farmers in the region.

Keywords: Drought stress, Growth, Tropical alfalfa

1 Introduction

The leguminous plant known as alfalfa (*Medicago sativa* L.) has excellent potential for use as a premium animal feed ingredient. This plant is well known for having a high crude protein content, readily digested fiber, and several vital vitamins and minerals [1]. However, as a plant that typically thrives in temperate regions, adapting alfalfa to tropical environments poses a unique challenge, particularly in the context of increasingly intense global climate change.

Alfalfa, which has begun to be cultivated in tropical regions [2], shows superior growth and nutrient content and can compete with subtropical varieties, creating opportunities for its recognition as a new tropical alfalfa variety [3]. The Indonesian Ministry of Agriculture officially recognized this variety, Kacang Ratu BW, registered with the Plant Variety Protection and Agricultural Licensing Center under number 929/PVHP/2021. Research on tropical alfalfa (*Medicago sativa* cv. Kacang Ratu BW) has significantly progressed since 2010 [4].

* Corresponding author: sriwidiastuti@mail.ugm.ac.id

Climate change has increased the frequency and intensity of droughts in various regions, including tropical areas such as Indonesia [5]. This makes drought stress one of the main limiting factors in the growth and productivity of alfalfa in tropical regions. Therefore, understanding the response of tropical alfalfa to various levels of drought stress becomes crucial to optimizing the cultivation of this plant in facing climate change challenges.

This study aims to investigate the growth of tropical alfalfa in response to different levels of drought stress, represented by varying irrigation frequencies. Additionally, this research observes the impact of cutting age on the plant's growth. In the context of livestock production, this research holds significant importance. Alfalfa is a crucial component in livestock production systems [3]. The availability of high-quality forage throughout the year is an important challenge for farmers, especially in tropical areas prone to climatic fluctuations [6]. By understanding the response of tropical alfalfa to drought stress, this research can provide a scientific basis for developing more effective management strategies for alfalfa cultivation in tropical regions.

2 Materials and Methods

2.1 Materials

The study was conducted over three months, from September to November 2023, at the Forage and Pasture Research Field, Faculty of Animal Science, Universitas Gadjah Mada, Yogyakarta, Indonesia. Field equipment included a 5 m measuring tape for plot and plant height measurements, soil preparation tools (sickle, hoe, tiller), and hoses for plant watering. The study materials comprised regrowth-sixth tropical alfalfa plants, regosol soil, and water.

2.2 Methodology

Plant growth measurements were conducted weekly using a Nested Design. The first factor was irrigation frequency with watering intervals of 1 day (P0) as a control, 2 days (P1), 4 days (P2), and 6 days (P3). The second factor was the cutting age after the sixth regrowth, consisting of 20-day and 30-day cutting ages. The observed variables included plant height, growth rate, and branch count.

Plant maintenance activities included watering to field capacity, weeding, and pest and disease control. Watering to field capacity meant providing water until the soil was saturated, with excess water allowed to drain. Field capacity refers to the maximum amount of water soil can hold against gravitational pull. Weeding was carried out regularly to clear weeds and loosen the soil carefully to avoid damaging plant roots. Pest and disease control were performed mechanically by removing pests like caterpillars and grasshoppers, which only attacked a few plants.

2.3 Measured variables

2.3.1 Plant height

Plant height was measured from the plant base at soil level to the tip of the plant using a ruler in centimeters (cm). This measurement aimed to gauge plant growth vigor and development under environmental conditions.

2.3.2 Growth rate

The growth rate refers to the speed at which a plant grows over a specific period. It is measured through changes in plant height, leaf count, plant mass, or other growth parameters over a given interval, usually expressed in growth units per day or week.

2.3.3 Number of branches

The number of branches was determined by counting the productive lateral branches, which can produce shoots or flowers and contribute to overall plant productivity.

2.4 Data analysis

The data obtained were statistically analyzed using Nested Design analysis, followed by Duncan's Multiple Range Test (DMRT) if a significant difference was observed at a confidence level of $\alpha = 0.05$, using SPSS version 26.

3 Results and Discussion

3.1 Agroclimatic conditions

Based on data from the Yogyakarta Climatology Station, climate conditions from September to November 2023 at the research location indicated average temperatures ranging from 18.4°C to 35.6°C with humidity between 66-84%. Rainfall from September to early November was recorded at 0 mm/month. The monthly average duration of sunshine reached 7.5 hours. This aligns with previous research that indicated optimal subtropical alfalfa growth at temperatures around 20°C, while tropical alfalfa can thrive even at temperatures reaching 48°C [15].

3.2 Plant height

Plant height is a growth indicator used to measure the response of plants to the treatments applied in an experiment. It represents a variable that can indicate the vegetative growth activity of a plant. The height of tropical alfalfa under different drought stresses is shown in Table 1.

Table 1. Average plant height of tropical alfalfa under different drought stresses (cm)

Irrigation Frequency	Cutting Age	
	20 days	30 days
P0	59.78±7.51	72.78±3.38
P1	55.89±4.59	69.67±3.54
P2	57.61±6.84	70.33±2.59
P3	61.61±6.07	69.67±4.42

Note: P0, Daily irrigation; P1, Irrigation every 2 days; P2, Irrigation every 4 days; P3, Irrigation every 6 days.

Analysis revealed that different cutting ages and irrigation frequencies, such as P0, P1, P2, and P3, did not significantly ($P>0.05$) affect the plant height of tropical alfalfa under drought stress (Table 2). This result is higher than the previous research [7], which reported alfalfa height of 44.78 cm at 60 days of age and 50.63 cm at 80 days of age. Furthermore, previous studies [3] showed that tropical alfalfa harvested at 3 weeks and 4 weeks of age reached heights of 10.42 cm and 30.11 cm, respectively.

This occurs because alfalfa has adaptive mechanisms that enable the plant to survive even under drought conditions. Alfalfa has a deep root system and efficient water use, enabling it to access water from deeper soil layers even when irrigation is not performed regularly [6]. Alfalfa roots typically grow deeper than other plants, reaching up to 16 feet (4.87 m) into the soil [8]. This deep root system creates channels that help retain soil, increase water infiltration and biological activity in the root zone, and improve nutrient cycling. Additionally, alfalfa exhibits drought tolerance due to its ability to reduce evaporation and increase water use efficiency under dry conditions [9]. Plants that have undergone six regrowths have deeper roots, making them more resilient to drought stress. The depth of alfalfa roots increases linearly with age, from 3 to 12 years, with an average increase of 35,3 cm per year. Based on this data, the root depth of alfalfa at 10, 20, and 30 years is estimated to reach 613 cm, 966 cm, and 1319 cm, respectively [10].

3.3 Growth rate

The growth rate of alfalfa refers to the speed at which the alfalfa plant grows. The growth rate data in this study is shown in Figure 1.

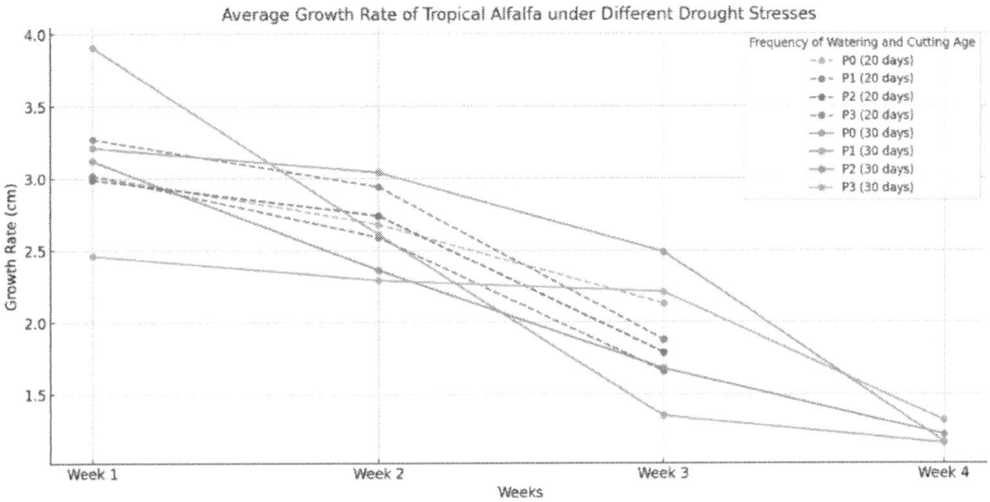

Note: P0, Daily irrigation; P1, Irrigation every 2 days; P2, Irrigation every 4 days; P3, Irrigation every 6 days.

Figure 1. Average growth rate of tropical alfalfa under different drought stresses

Based on the analysis, there were no significant differences in the growth rate of tropical alfalfa across different irrigation frequencies and cutting ages. Although there were variations in average plant height with each treatment, the overall decrease in height did not show statistically significant differences ($P>0.05$). The declining pattern across all treatments tended to be similar, where initial growth appeared promising in the first weeks but then gradually decreased over time, especially at longer cutting ages.

These results indicate that although there are slight variations in alfalfa's response to irrigation frequency under drought conditions, the overall impact on growth rate is not statistically significant. This suggests that tropical alfalfa has a relatively similar adaptive capacity in coping with drought stress regardless of the irrigation frequency. This aligns with previous findings [2], which noted that alfalfa's average height or growth rate in the first week after cutting increased between 2-5 cm per day. However, as the flowering process began, the plant's growth gradually declined in the third and fourth weeks.

3.4 Number of branches

The number of branches refers to the total number of branches formed from the main stem of a plant. These branches develop from buds that emerge on the main stem and contribute to the plant's overall growth and development. The number of branches in tropical alfalfa under different drought stresses is shown in Table 2.

Table 2. Average number of branches of tropical alfalfa under different drought stresses (count)

Irrigation Frequency	Cutting Age	
	20 days	30 days
P0	23.22±3.42	23.67±3.32
P1	20.89±2.52	25.11±3.85
P2	22.56±3.09	24.56±3.13
P3	24.44±2.60	25.33±3.24

Note: P0, Daily irrigation; P1, Irrigation every 2 days; P2, Irrigation every 4 days; P3, Irrigation every 6 days.

Based on the data above, no significant differences ($P>0.05$) were found in the number of branches of tropical alfalfa across different irrigation frequencies at different cutting ages. The data shows that the number of branches in P0 varied from 20.89±2.52 to 24.44±2.60 for the 20-day cutting age, while for the 30-day cutting age, the number of branches ranged from 23.67±3.32 to 25.33±3.24. Although other irrigation frequencies, such as P1, P2, and P3, showed different numbers of branches, the same trend occurred in all treatments, with no statistically significant changes observed.

This indicates that tropical alfalfa shows a similar response to drought stress conditions, without a meaningful effect of irrigation frequency on the number of branches produced. The process of branch formation in tropical alfalfa under different irrigation frequencies involves a series of complex physiological and morphological responses. Initially, alfalfa produces a primary shoot. As the plant grows, it develops apical and axillary meristems that have the potential to form new branches [11]. The plant can also divert resources from branch growth to develop deeper root systems in search of water. Additionally, plant hormones such as auxin and cytokinin play essential roles in regulating apical dominance and lateral branching, and the balance of these hormones can change in response to water stress [12].

4 Conclusion

This study shows that tropical alfalfa (*Medicago sativa* cv. Kacang Ratu BW) exhibited no significant differences in plant height, growth rate, and number of branches at various irrigation frequencies and cutting ages. This plant demonstrates good adaptability to drought stress. The results of this study indicate that tropical alfalfa has the potential to be cultivated in tropical environments that often experience drought and can provide knowledge in effective cultivation strategies in the tropics.

Acknowledgments

The authors wish to express their sincere gratitude to the Indonesia Endowment Fund for Education (Lembaga Pengelola Dana Pendidikan/LPDP) for providing financial support. This support has been instrumental in facilitating the completion of this paper.

References

1. M. Roy, J. Niu, A. Irshad, H.A. Kareem, M.U. Hassan, N. Xu, X. Sui, Z. Guo, A. Amo, Q. Wang, Exogenous melatonin protects alfalfa (*Medicago sativa* L.) seedlings from drought-induced damage by modulating reactive oxygen species metabolism, mineral balance and photosynthetic efficiency, Plant Stress. 2, 100044 (2021). https://doi.org/10.1016/j.stress.2021.100044
2. B. Suwignyo, F.X.D. Kurniawan, N. Suseno, R. Utomo, B. Suhartanto, Productivity and nutrient content of the second regrowth alfalfa (*Medicago sativa* L.) with different photoperiod and dolomite, Animal Production. 22(2), 74-81 (2020). https://doi.org/10.20884/1.jap.2020.22.2.53
3. B. Suwignyo, L. Arifin, N. Umami, Muhlisin, B. Suhartanto, The performance and genetic variation of first and second generation tropical alfalfa (*Medicago sativa*), Biodiversitas Journal of Biological Diversity. 22(6), 3265-3270 (2021). https://doi.org/10.13057/biodiv/d220631
4. B. Suwignyo, E.A. Rini, S. Helmiyati, The profile of tropical alfalfa in Indonesia: A review, Saudi Journal of Biological Sciences. 30, 103504 (2023). https://doi.org/10.1016/j.sjbs.2022.103504
5. N. Nandgude, T.P. Singh, S. Nandgude, M. Tiwari, Drought prediction: a comprehensive review of different drought prediction models and adopted technologies, Sustainability. 15(15), 11684 (2023). https://doi.org/10.3390/su151511684
6. K. Wang, J. Yan, R. Tanvir, L. Li, Y. Liu, W. Zhang, Improved forage quality and biomass yield of alfalfa (*Medicago sativa* L.) by Arabidopsis QQS orphan gene. Current Plant Biology. 35-36, 100295 (2023). https://doi.org/10.1016/j.cpb.2023.100295
7. R.D. Wahyuni, S.N. Kamaliyah, Study on production pattern of tropical alfalfa (*Medicago sativa* L.), Jurnal Ilmu-ilmu Peternakan. 19(1), 20-27 (2009). https://jiip.ub.ac.id/index.php/jiip/article/view/107
8. A. Fernandez, C. Sheaffer, N. Tautges, D. Putnam, M. Hunter, Alfalfa, Wildlife, and the Environment (2nd ed.), (National Alfalfa and Forage Alliance, St. Paul, 2019)
9. M. Liu, X. Wu, H. Yang, Evapotranspiration characteristics and soil water balance of alfalfa grasslands under regulated deficit irrigation in the inland arid area of Midwestern China, Agricultural Water Management. 260, 107316 (2022). https://doi.org/10.1016/j.agwat.2021.107316
10. P. Liu, W. Hao, J. Li, Z. Jia, Study on curve fitting features of soil moisture and root system's dynamic distribution in alfalfa grassland in drought areas of southern Ningxia, Journal of Hebei Agricultural University. 34(4), 29-34 (2011)
11. J.A.H. Murray, A. Jones, C. Godin, J. Traas, Systems analysis of shoot apical meristem growth and development: integrating hormonal and mechanical signaling, The Plant Cell. 24(10), 3907-3919 (2012). http://www.jstor.org/stable/23327031
12. E. Nambara, S.C.M. Van Wees, Plant hormone functions and interactions in biological systems, The Plant Journal. 105(2), 287–289 (2021). https://doi.org/10.1111/tpj.15151

Effect of zero-valent iron on *Rhizobium* sp. cells isolated from cadmium-contaminated sites after remediation by zero-valent iron

Aussanee Nuiyen[1], *Vinta*-Khum-in[2], and *Siriwan* Wichai[1*]

[1] Department of Microbiology and Parasitology, Faculty of Medical Science, Naresuan University, Phitsanulok, Thailand

[2] Division of Environmental Engineering, Department of Civil Engineering, Faculty of Engineering, Rajamangala University of Technology Thanyaburi (RMUTT), Pathum Thani, Thailand

Abstract. Cadmium contamination found in paddy fields in the Maesot District of Tak Province, Thailand. This area was remediated using 50mg/L of ZVI. The study aimed to isolate and identify soil bacteria in the soil and rice roots and to investigate ZVI's effect on the isolated bacterial cells. The results indicated no significant difference in soil bacteria content before and after remediation at the 95% confidence level. Twelve isolates of nitrogen-fixing bacteria were obtained. Those isolates could grow at high concentrations of 300 mg/L of ZVI. RH17 had a high tolerance for TSA with 300 mg/L of ZVI at only 10 CFU/ml. The effects of ZVI at 150 mg/L on RH17 cells, a small amount of ZVI was observed adhering to the cells' surface and forming giant cells, while at 300 mg/L of ZVI, caused a reduction in growth by 81.0%. The *nif*H gene of RH17 was related to *Rhizobium* sp. strain 5-1-2. The results demonstrated the cadmium remediation process with 50mg/L of ZVI did not affect the cell count of soil bacteria in the paddy field. However, at 150 mg/L or higher, ZVI damaged the isolated *Rhizobium* sp. cell membrane. So, the remediation using ZVI must consider the appropriate concentration.

1. Introduction

Toxic and heavy metal contamination in the environment is a long-standing problem that harms humans and the environment. Cadmium contamination in paddy fields in the Maesot District of Tak Province, Thailand, is caused by improper mining practices and inadequate waste management [1]. In the case being studied, it is important to note that there was no decrease in cadmium bioaccessibility over time. Even after four years of cadmium aging in the soil, the bioaccessibility remained very high, accounting for about 60% of the total content in most contaminated soil samples [2]. Traditional techniques, such as excavation and landfilling, are expensive and unstainable for removing contaminated soil. However, adding immobilizing agents to contaminated soil can reduce metal mobility and bioavailability, making soil amendment cost-effective and less disruptive [3]. One effective agent for reducing heavy metal contamination in soil is zero valent iron (ZVI). It is a low-

* Corresponding author: siriwanwichai@nu.ac.th

cost amendment that can be used for both reductive and sorptive sequestration of heavy metals [4,5]. ZVI can effectively remediate cadmium-contaminated soil by binding directly to cadmium in soil. Stevenson et al [6] found that iron's ability to bind to cadmium charges effectively reduces cadmium levels in water by up to 80%. However, it should be noted that the ZVI that binds to cadmium can also increase toxicity and significantly reduce microbial activity [7].

Iron is a practical and effective method to bind metal in soil, as stated by Goncalves [8]. However, oxidative stress or the Fenton reaction can adversely affect this method by damaging cell membranes and genetic material. This can result in protein and lipid oxidation, leading to microorganism death or cell rupture [9]. Moreover, short-term exposure to Nanoscale ZVI (nZVI) can reduce freshwater microalgal growth and impact the aquatic environment, as observed in the research by D´ors et al. [10].

In 2020, Khum-in et al [11] successfully reduced cadmium concentration in the Phatat Pha Daeng Subdistrict Mae Sot District, Tak Province, Thailand. They achieved this by using a combination of biochar and zero-valent iron. Community researchers found that the application of biochar with zero-valent iron (BZVI) as a soil amendment reduced cadmium and enriched Fe and Zn in rice grains. In a later study 2023, Khum-in et al [12] indicated that magnet-assisted soil washing followed by soil amendment using ZVI also successfully removed cadmium from soil and reduced cadmium content in rice grains. However, the effects of this remediation process on soil microorganisms have not yet been explored. In rice paddy soil, biological nitrogen fixation is important for sustaining soil nitrogen fertility and rice growth [13]. This study aims to isolate and identify nitrogen-fixing bacteria from soil and rice roots in a cadmium-contaminated area remediated with 50mg/L ZVI. The study also investigates the effects of ZVI on the growth of obtained isolates.

2. Materials and methods

2.1 Isolation of soil bacteria related to nitrogen-fixing bacteria and phenotypic characterization

2.1.1 Bacteria isolation and enumeration

10 grams of soil contaminated with cadmium and remediated using ZVI were placed in a 250 mL Erlenmeyer flask containing 90 mL of 0.85% NaCl. The control soil samples were taken from the paddy fields before remediation. The mixture was shaken at 120 rpm for 30 minutes. A serial dilution was performed by adding 0.85% NaCl to a dilution of 10^{-5}. The dilution of 10^{-3} to 10^{-5} was then spread onto plates containing Yeast mannitol agar with Congo red (Himedia). These plates were incubated for 2-5 days at 30°C to observe colonies of bacteria and enumeration. After remediation with 50mg/L of ZVI, the rice was grown for 12 weeks. The method used for isolating nitrogen-fixing bacteria from rice roots was based on Mbai FN et al. [14]. The morphology of different colonies was isolated and subcultured in the same media for further study. The number of bacteria before and after remediation were compared at the 95% confidence level.

2.1.2 Nitrogen fixing activity

Isolated bacteria were cultured in a nitrogen-free mineral medium at 30°C to evaluate nitrogen-fixing activity. The medium's color change was recorded after incubation for 3-7 days. Afterward, the ammonium test kit reagents were added, and the resulting color was recorded by comparing it to the color chart provided [15].

2.2 The effect of ZVI on bacterial cells

2.2.1 Survival of bacterial cells in the presence of different concentrations of ZVI

Isolated bacteria were cultured in Tryptic soy broth and then incubated at 30°C for 48 h. Afterward, centrifugation was performed at 6,000 rpm for 5 min. The cells were washed twice with 150 mM phosphate-buffered saline (pH 7.2) and mixed with 50 mM carbonate buffer (pH 8.0). The concentration of cells was adjusted to approximately $1.0x10^8$-$3.0x10^8$ CFU/ml in 25 mL carbonate buffer. These cells were added to tubes containing 50, 100, 150, 200, 250, and 300 mg/L ZVI and then incubated at 30°C for 48 h [16]. The viability of the bacteria was evaluated using a spread plate technique on TSA agar, which was then incubated at 30°C for 48 h.

2.2.2 Survival of bacterial cells with varying concentrations in the presence of 300 mg/L of ZVI

Isolated bacteria were cultured in Tryptic soy broth and then incubated at 30^0C for 48 h. The cell concentration was adjusted to approximately $1.0x10^8$-$3.0x10^8$ CFU/ml in 25 mL carbonate buffer and then diluted to different cell concentrations (CFU/mL). A variable number of cells with different concentrations were added to the TSA medium with 300 mg/L ZVI. Then, the plates were incubated at 30°C for 24 h, and bacterial growth was observed [17].

2.2.3 The effects of ZVI on bacterial cells via Transmission Electron Microscopy

Isolated bacteria were cultured in Tryptic soy broth and then incubated at 30°C for 48 h. Afterward, centrifugation was performed at 6,000 rpm for 5 min. The cells were washed twice with 0.1M ammonium acetate. The concentration of cells was adjusted to approximately $1.0x10^8$- $3.0x10^8$ CFU/mL. Cells were diluted in 0.1M ammonium acetate to 10^4 CFU/mL, treated with 150 and 300 mg/L ZVI, and then shaken at 180 rpm for 60 minutes to distribute the ZVI throughout the solution. The survival of bacteria was checked using the spread plate technique on a TSA medium, and the results were compared with those of the control. Some samples were prepared for study under a Transmission Electron Microscope using negative staining with uranium acetate [18]. The altered cell characteristics were studied at the Science Lab Center, Faculty of Science, Naresuan University.

2.3 Phylogenetic analysis of *nifH* gene

2.3.1 DNA extraction, purification, and amplification

The selected bacteria were grown in Tryptic soy broth under shaking at 30°C for 18 h. The resulting bacterial suspension was then centrifuged at 12,000xg for 12 min. The centrifuged sample was further processed using the Genomic DNA extraction Mini kit, and the extracted DNA was stored at -20°C for further study [19]. DNA amplification with *nif*H genes, approximately 760 bp of *nif*H gene was amplified by using *nif*H1 5′ CTG TTT ACG GCA AGC GCG GTA TCG GCA 3′ as forward primer and *nif*H2 5′TCC TCC AGC TCC TCC ATG GTG ATC GG 3′ as reverse primer in 50 µL PCR Master mix under DNA amplification conditions. The reaction conditions were an initial denaturation at 95°C for 5 min, followed by 35 cycles of denaturation at 94°C for 60 s, annealing at 55°C for 30 s, extension at 72°C for 60 s, and a final extension at 72°C for 10 min. PCR products were stored at 4°C. Ten microliters of PCR product were examined using gel electrophoresis at a 1.2% agarose gel concentration in 0.5X TBE Buffer, and then the DNA band was measured using UV Light Transilluminators [20].

2.3.2 DNA sequencing

The purified PCR product was sequenced by Macrogen Korea and aligned using GenBank's BLAST function and the National Center Biotechnology information on the NCBI website. The data set's evolution tree was inferred using the Maximum Likelihood method based on the Kimura-2-parameter model. Evolution analyses were conducted using Evolutionary Genetics Analysis (MEGA) software version 7 [21].

2.4 Statistical analysis

The data analysis was performed by Stata (Stata12.0 Corporation, USA). The One-Way ANOVA was analyzed by comparing multiple groups of LogCFU/g before and after remediation. The P-values < 0.005 were compared with the Greenhouse Geisser epsilon, considered statistically significant.

3. Results

3.1 Isolation of soil bacteria related to nitrogen-fixing bacteria and phenotypic characterization

The levels of soil bacteria related to nitrogen-fixing bacteria in the soil before and after remediation with ZVI were compared. The results showed that the number of bacteria before remediation was 6.33 log, which was not significantly different from after remediation (6.95 log) at the 95% confidence level. Fifty-six bacterial isolates were found on YMA plates with Congo red, with forty-four originating from soil and eight from roots. Among these, sixteen were identified as gram-positive and forty as gram-negative bacteria. Microscopic analysis revealed that fifty-one isolates had rod-shaped cells, while three had filamentous shapes, and two had rod shapes with rounded ends. From the catalase and cytochrome oxidase synthesis tests, it was found that fifty-two bacterial isolates exhibited catalase synthesis, while only four isolates showed negative results. The cytochrome oxidase test showed positive results for thirty-five isolates and negative results for twenty-one isolates (Table 1).

Table 1. Morphological and biochemical characteristics of 56 bacteria isolated from soil and root samples.

Samples	Gram's		Shape			Catalase		Oxidase	
	+	-	rod	bacilli	hyphae	+	-	+	-
Soil	12	36	44	2	2	44	4	28	20

Rice root	4	4	7	-	1	8	-	7	1
Total	16	40	51	2	3	52	4	35	21

Twelve bacterial isolates had biochemical properties similar to those of nitrogen-fixing bacteria, *Rhizobium* sp. Out of these twelve, ten were found in soil (RH5, RH6, RH12, RH17, RH22, RH23, RH25, RH29, RH34, and RH42), and two were found in roots (RH2R and RH7R) (Table 2).

Table 2. The morphology and biochemistry of isolated bacteria were compared to *Rhizobium* sp. to determine the amount of fixable nitrogen. * The study conducted by Yu et al., 2011 [15].

Isolates	Morphological and biochemical test				The amount of fixable nitrogen
	Gram's	Cell shape	Catalase	Oxidase	
Rhizobium spp. *	-	Rod	+	+	2*
RH5	-	Rod	+	+	<0.25
RH6	-	Rod	+	+	1.50
RH12	-	Rod	+	+	<0.25
RH17	-	Rod	+	+	<0.25
RH22	-	Rod	+	+	0.25
RH23	-	Rod	+	+	<0.25
RH25	-	Rod	+	+	0.25
RH29	-	Rod	+	+	0.25
RH34	-	Rod	+	+	<0.25
RH42	-	Rod	+	+	<0.25
RH2R	-	Rod	+	+	<0.25
RH7R	-	Rod	+	+	0.25

3.2 The effect of ZVI on bacterial cells

All twelve isolates that gave similar physiological and biochemical test results to bacteria in nitrogen-fixing bacteria were tested for resistance to ZVI at concentrations of 50-300 mg/L. After conducting the test, it was discovered that all twelve isolates survived at the maximum concentration of 300mg/L ZVI, even when their cell concentration was 1.0×10^8 CFU/mL. This was in contrast to *Escherichia coli* DH5α and *Escherichia coli* ATCC25922, which are sensitive to heavy metals and antibiotics and could only survive at a maximum ZVI concentration of 300mg/L.

The survival of bacterial cells with varying concentrations when exposed to ZVI at 300 mg/L. This study discovered that ZVI could prevent the growth of bacteria. Among the several bacteria tested, three isolates (RH5, RH6, and RH2R) could withstand a concentration of 300 mg/L of ZVI at a cell concentration of 10^3 CFU/mL. Moreover, at a cell concentration of 10 CFU/mL, four other isolates (RH17, RH22, RH25, and RH34) exhibited tolerance to ZVI (Table 3). Out of four isolates, RH17 was identified as the fastest grower. Consequently, RH17 was chosen for further studies.

Table 3. The concentration of bacterial cells that can be grown on TSA medium with 300mg/L of ZVI.

Isolates	Cell concentrations (CFU/mL)
Escherichia coli DH5α	10^4
Escherichia coli ATCC25922	10^4

RH6, RH7, RH2R	10^3
RH12, RH23, RH29, RH42, RH7R	10^2
RH17, RH22, RH25, RH34	10

The effects of ZVI on bacterial cells were analyzed via Transmission electron microscopy. RH17 was grown in a Tryptic soy broth with 150 mg/L of ZVI. After being shaken at 180 rpm for 60 minutes, the growth decreased by 71.16%. A slight adhesion of ZVI to the cell surface increased cell size, leading to the formation of more giant cells compared to the control group. In contrast, the growth increased by 13% in the control group, and the cells remained in good condition. However, their cell membranes began to deteriorate when bacterial cells were grown in a TSB medium with 300 mg/L of ZVI. ZVI was found to stick to the cell membrane, and the growth decreased by 81.0% (Table 4, Figure 1).

Table 4. Growth of isolated bacteria RH17 in TSB medium with and without added ZVI after 60 minutes of incubation.

ZVI conc. (mg/L)	Cell Number (CFU/mL)		Percent (%)	
	Control	After 60 min	Increased	Decreased
0	1.00×10^4	1.13×10^4	13	-
150	1.00×10^4	2.83×10^3	-	71.76
300	1.00×10^4	1.90×10^3	-	81.00

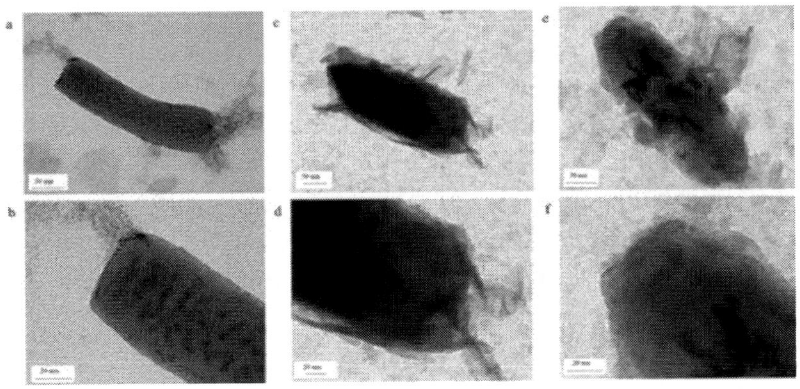

Fig. 1. shows the bacterial cells of the RH17 observed under a Transmission electron microscope (TEM) after being grown for 60 minutes in a TSB medium with and without the addition of zero-valent iron.

a-b Characteristics of bacterial cells in TSB medium

c-d Characteristics of bacterial cells in TSB medium containing 150 mg/L ZVI

e-f Characteristics of bacterial cells in TSB medium containing 300 mg/L ZVI

3.3 Phylogenetic analysis of *nif*H gene

After amplifying *nif*H genes through a polymerase chain reaction, three isolates (RH12, RH17, and RH2R) displayed a genetic material band of approximately 760 bases (Figure 2). To evaluate the phylogenetic analysis of *nif*H gene, the purified product was sequenced by Macrogen Korea. The nucleotide sequence was aligned using the BLAST function of GenBank and the National Center Biotechnology information on the NCBI website (http://ncbi.nlm.nih.gov). The data set was inferred using the Maximum Likelihood method based on the Kimura-2 parameter model. Evolution analyses were conducted using

 https://doi.org/10.1051/bioconf/202516403001

Evolutionary Genetic Analysis (MEGA) software version 7 [21]. The phylogenetic analysis of the *nif*H gene RH17 is shown in Figure 3.

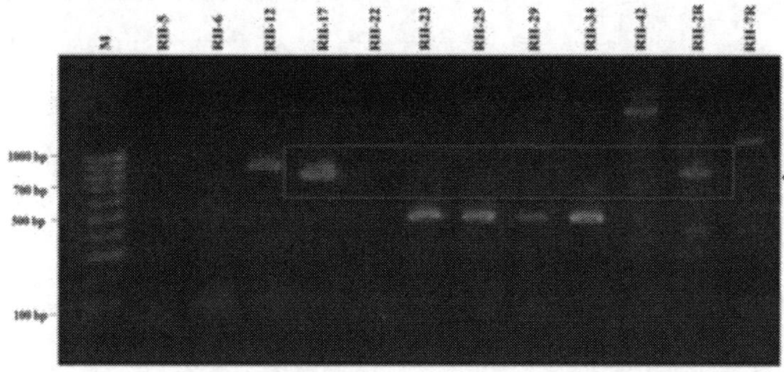

Fig. 2. The band of genetic material that appears after amplifying the *nif*H gene through polymerase chain reaction.

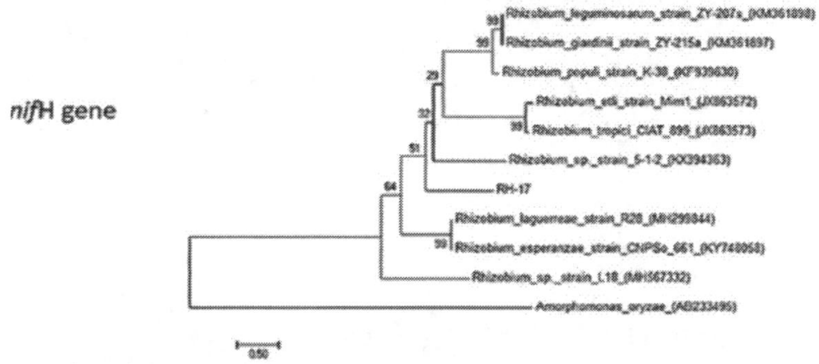

Fig. 3. Molecular phylogenetic analysis by Maximum Likelihood method. The Maximum Likelihood method based on the Kimura-2 model inferred the evolutionary history. The three with the highest log likelihood (-6645.8643) are shown. The percentage of trees in which the associated taxa clustered is shown next to the branches. The heuristic search's initial tree (s) were obtained automatically using the Maximum Parsimony method. The tree is drawn to scale, with branch lengths measured in the number of substitutions per site. The analysis involved 11 nucleotide sequences. Codon positions included were 1st+2nd+ 3rd+noncoding. All positions containing gaps and missing data were eliminated. There was a total of 626 positions in the final dataset. Evolutionary analyses were conducted in MEGA7.

4. Discussion

The amounts of soil bacteria related to nitrogen-fixing bacteria before and after soil remediation with ZVI 50 mg/L in the Phatat Pha Daeng Mae Sot District, Tak Province, Thailand, were not significantly different. Therefore, bacteria can grow and multiply at 50 mg/L ZVI. So, it is possible to conclude that ZVI at this concentration did not affect the growth of soil bacteria. This could be explained by the fact that these isolated bacteria had a high tolerance to ZVI because, in an appropriate environment, they have a high cadmium concentration. As a result, bacteria have adapted to survive in such conditions by using mechanisms such as expelling metals from the cell or reducing the uptake of metals into the cell [22].

The isolation of fifty-six bacterial isolates with varying colonies resulted from the soil and rice roots in the cadmium remediation area: forty-eight from the soil and eight from rice roots. Fifty-six bacterial isolates' physiological and biochemical characteristics revealed twelve shared characteristics with the nitrogen-fixing bacteria. Of the twelve isolates tested, five could fix nitrogen. These five isolates were RH6, RH22, RH25, RH29, and RH7R. They can fix nitrogen within the 0.25-1.50 mg/L range. This site in the Maesot District, Tak Province, Thailand, was remediated using magnet-assisted soil washing and soil amendment with 50mg/L ZVI to remove cadmium contamination. After 12 weeks of remediation, the concentration of cadmium in the soil decreased from 41.30±3.50 mg/kg to 34.82±6.27 mg/kg, while the concentration of iron increased from 33.49±0.60 mg/kg to 54.71±1.75 mg/kg [10]. The presence of ZVI during the remediation process may impact the nitrogen-fixing bacteria in the area. Therefore, the next step was to assess the impact of higher iron content on these bacteria.

The effect of ZVI on bacterial cells: It has been found that all twelve bacterial isolates, namely RH5, RH6, RH12, RH17, RH22, RH23, RH25, RH29, RH34, RH42, RH2R and RH7R, can survive at a ZVI concentration of 300 mg/L, even when their cell concentration is $1.0x10^8$ CFU/mL. Among the bacteria tested, three isolates (RH5, RH6, and RH2R) were able to withstand a concentration of 300 mg/L of ZVI at a cell concentration of $1.0x10^3$ CFU/mL. In contrast, four isolates (RH17, RH22, RH25, and RH34) could grow at a cell concentration of 10 CFU/mL. It has been shown that these bacteria were highly tolerant to ZVI.

After examining bacterial cells of RH17 using a Transmission electron microscope in a Tryptic soy broth medium with 150 mg/L of ZVI, there was a slight adhesion of ZVI on the cell surface. This increased in cell size, leading to the formation of more giant cells compared to the control group. It was observed that the growth decreased by 71.16%. Bacteria can protect themselves against damage caused by heavy metals in several ways. They can produce proteins that bind to metals outside the cell membrane, excrete heavy metals from cells through efflux transporters, synthesize metal-binding proteins within cells to reduce metal toxicity or create organic substances outside the cell to prevent heavy metals from entering cells. This mechanism enables bacteria to thrive in an environment with high levels of metal [23].

Moreover, it was observed that the characteristics of RH17 bacterial cells were altered when the concentration of ZVI was increased to 300 mg/L. Direct contact with ZVI caused deterioration of the bacterial cell wall. ZVI directly affects the phospholipid fatty acid (PLFA) layer around bacterial genetic material. From the report of cell membrane composition analysis, a study reported that when *Pseudomonas putida* F1 is cocultured with ZVI nanoparticles, there is a change in cis-undersaturated fatty acid to trans isomers when the bacterial cells come into contact with the nanoparticles. This makes the bacterial cell membrane stiffer in the short term instead of being soft [24]. Reports have suggested that entering Fe(II) iron into *Escherichia coli* cells can damage their cell membrane. This can trigger the generation of reactive oxygen species (ROS), such as oxygen or hydrogen peroxide, and ultimately lead to oxidative stress. As a result, the bacterial cells may rupture [25].

The results show that the isolate RH17 was highly tolerant to ZVI, which is different from the research conducted by Chen et al. (2013) [26]. Their study investigated the effect of ZVI on the growth of denitrifying bacteria and the denitrification process. Their findings revealed that only high concentrations of ZVI impacted the growth of microorganisms. When the concentration exceeded 20 mg/L, it affected the growth and activities of bacteria, particularly during the denitrifying process.

Phylogenetic analysis of the *nif*H gene revealed that PCR products of approximately 760 bases were expressed by three nitrogen-fixing bacteria isolates, RH12, RH17, and RH2R. The *nif* gene plays an important role in nitrogenase enzymes found in bacteria and archaea, as they convert nitrogen into ammonia through nitrogen fixation. This gene can be found in many bacteria, such as *Klebsiella pneumonia*, *Azotobacter vinelandii*, and *Pseudomonas stutzeri* [27]. However, some bacterial isolates could fix nitrogen without expressing *nif*H, a nitrogenase Fe protein synthesis gene. This gene is commonly found in Proteobacteria,

Cyanobacteria, Firmicutes, and Actinobacteria [28]. It is possible for bacterial isolates without *nif*H genes to still fix nitrogen due to the presence of other *nif* genes, such as *nif*D and *nif*K [29]. Some bacteria have different combinations of *nif* genes. For example, *Rhodobacter capsulatus*, has *nif*AB operon and *nif*U, while *Azotobacter vinelandii* has *nif*E and *nif*N. It has been reported that *Rhizobium* sp.NGR 234 is missing the *nif*H gene. However, the bacterial isolates RH5, RH6, RH7R, RH22, RH23, RH25, RH29, RH34, and RH42 may use *nif* genes other than *nif*H to fix nitrogen [30].

Three nitrogen-fixing bacteria isolates, RH12, RH17, and RH2R, had the *nif*H gene, and RH17 was highly tolerant to ZVI. After phylogenetic analysis of the *nif*H gene, RH17 was identified as belonging to *Rhizobium* sp. strain 5-1-2 (GenBank Accession Number KX394363.1).

5. Conclusions

Twelve isolates of nitrogen-fixing bacteria were obtained from the soil and rice roots in Maesot District, Tak Province, Thailand, which had undergone remediation cadmium by using 50mg/L of ZVI. These isolates could grow at high concentrations of 300 mg/L of ZVI. After exposing RH17 cells to a TSB containing 150 mg/L of ZVI, a small amount of ZVI was observed adhering to the cells' surface and forming giant cells, while at 300 mg/L of ZVI, the deterioration of their cell membrane caused a reduction in growth by 81.0%. The *nif*H gene of RH17 was related to *Rhizobium* sp. strain 5-1-2. Our findings revealed that nitrogen-fixing bacteria have a high tolerance to ZVI, necessitating further research into gene identification and the functional characterization of the nitrogen cycle.

Acknowledgments: This work was financially supported by the National Research Council of Thailand (Number 102/2559).

References

1. T. Phenrat, P.Hongkumnerd, J. Suk-in, and V.Khum-in, Nanoscale zerovalent Iron particles for magnet-assisted soil washing of cadmium-contaminated paddy soil: proof of concept. Environ.Chem. **16**, 6 (2019). https://doi.org/10.1071/EN19028
2. G. Petruzzelli, M. Barbafieri, E. Franchi, D. Fusini, M. Vocciante, and F. Pedron, Effects of soil aging on cadmium bioavailability and bioaccessibility at a contaminated site. Environments. **10**, 105 (2023). https://doi.org/10.3390/environments10060105
3. Y. Hamid, L.Tang, M.I. Sohail, X. Cao, B. Hussain, and M.Z. Aziz, An explanation of soil amendments to reduce cadmium phytoavailability and transfer to food chain. Sci. Total Environ. **660**, (2019). https://doi.org/10.1016/j.scitotenv.2018.12.419
4. J.T. Qiao, T.X. Liu, X.Q.Wang, F.B. Li, Y.H. Lv, and J.H. Cui, Simultaneous alleviation of cadmium and arsenic accumulation in rice by applying zero-valent iron and biochar to contaminated paddy soils. Chemosphere. **195**, (2018). https://doi.org/10.1016/j.chemosphere.2017.12.081
5. T. Phenrat, and G.V. (Eds.) Lowry, Nanoscale Zerovalent Iron Particles for Environmental Restoration: From Fundamental Science to Field Scale Engineering Applications (Springer International Publishing, Switzerland, 2019b)
6. L. Stevenson, M. Adeleye, Y. Su, Y. Zhang, A. Keller, and R. M. Nisbet, Remediation of Cadmium Toxicity by Sulfidized Nano-Iron: The Importance of Organic Material. *ACS Nano*. **11**, 10 (2017). https://doi.org/10.1021/acsnano.7b05970
7. C. Fajardo, L.T. Ortiz, M.L.Rodríguez-Membibre, M. Nande, M.C. Lobo, and M. Martin, Assessing the impact of zero-valent iron (ZVI) nanotechnology on soil microbial structure and functionality: A molecular approach. Chemosphere. **86**, 8 (2012). https://doi.org/10.1016/j.chemosphere.2011.11.041
8. J. R. Gonçalves, The Soil and Groundwater Remediation with Zero Valent Iron Nanoparticles. Procedia Eng. **143**, (2016). https://doi.org/10.1016/j.proeng.2016.06.122
9. A. Ševcu, S. E. T. Yehia, E. J. Joner, and A. Cerník, Oxidative stress induced in microorganisms by zero-valent iron nanoparticles. M&E. **26**, 4 (2011). https://doi.org/10.1264/jsme2.ME11126
10. A. D'ors, A. Sanchez-Fortún, A. Cort´es-T´ellez, C. Fajardo, G. Mengs, M. Nande, C. Martín, G. Costa, M. Martin, M.C. Bartolom´e, and S. Sanchez-Fortún, Adverse effects of iron-based nanoparticles on freshwater phytoplankton *Scenedesmus armatus* and *Microcystis aeruginosa* strains. Chemosphere. **339**, 139710 (2023). https://doi.org/10.1016/j.chemosphere.2023.139710

11. V. Khum-in, J.Suk-in, P. In-ai, K. Piaowana, Y. Phaimisap, W. Supanpaiboon, and T. Phenrat, Combining biochar and zerovalent iron (BZVI) as a paddy field soil amendment for heavy cadmium (Cd) contamination decreases Cd but increases zinc and iron concentrations in rice grains: a field-scale evaluation. PSEP. **141**, (2020). https://doi.org/10.1016/j.psep.2020.05.008

12. V. Khum-in, J. Suk-in, P. In-ai, K. Piaowan, Y. Praimeesub, K. Rintachai, W. Supanpaiboon, and T. Phenrat, Combining magnet-assisted soil washing and soil amendment with zero-valent iron to restore safe rice cultivation in real cadmium-contaminated paddy fields. Chemosphere. **340**, 139816 (2023). https://doi.org/10.1016/j.chemosphere.2023.139816

13. Z. Zhengcheng, M. Yoko, X. Zhenxing, S. Yutaka, O. Hirotomo, and S. SKeishi, Active Nitrogen fixation by iron-reducing bacteria in rice paddy soil and its further enhancement by iron application. Applied Sci. **13**, 8156 (2023). https://doi.org/10.3390/app13148156

14. F.N. Mbai, E.N. Magiri, V.N. Matiru, J. Nganga, and V.C.S. Nyambati, Isolation and characterization of bacterial root endophytes with potential to enhance plant growth from Kenyan Basmati rice. Am.Int.J.Contemp.Res. **3**, (2013). https://repository.dkut.ac.ke:8080/xmlui/handle/123456789/7779

15. S. S. Yu, Z. K. Z. K., Latt, E. P. Kyaw, and T. M. Ly, Accumulation of ammonia in culture broth by wild-type nitrogen-fixing bacterium, *Stenotrophomonas maltophilia*. IJABPT. **2**, 1 (2011). https://imsear.searo.who.int/handle/123456789/161303

16. F. Kafilzadeh, Y. Moghtaderi, and A.R. Jahromi, Isolation and identification of cadmium-resistant bacteria in Soltan Abad river sediments and determination of tolerance of bacteria through MIC and MBC. Eur. J. Exp. Biol. **3,** 5 (2013).

17. S. A. Sabry, H. A. Ghozlan, and D. M. Abou-Zeid, Metal tolerance and antibiotic resistance patterns of a bacterial population isolated from sea water. J. Appl. Microbiol. **82,** 2 (1997). https://doi.org/10.1111/j.1365-2672.1997.tb03580.x

18. D. P. Pires, A. Dötsch, E. M. Anderson, Y. Hao, C. M. Khursigara, J. S. Lam, and J. Azeredo, A Genotypic Analysis of Five *P. aeruginosa* Strains after Biofilm Infection by Phages Targeting Different Cell Surface Receptors. Front. Microbiol. **8**, (2017). https://doi.org/10.3389/fmicb.2017.01229

19. D. S. Wagh, R. N. Shermale, and B. V. Mahure, Isolation and Characterization of Nitrogen Fixing Bacteria from Agricultural Rhizosphere. IOSR-JAVS. **8**, 6 (2015). DOI: 10.9790/2380-08624852

20. R. C. Dubey, D.K. Maheshwari, H. Kumar, and K. Choure, Assessment of diversity and plant growth promoting attributes of rhizobia isolated from *Cajanus cajan* L. Afr. J. Biotechnol. **9**, (2010). DOI:10.4314/AJB.V9I50

21. S. Kumar, G. Stecher, and K. Tamura, Molecular Evolutionary Genetics Analysis version 7.0 for bigger datasets. Mol. Biol. Evol.. **33**, 1870-1874 (2016). DOI: 10.1093/molbev/msw054

22. I.J. Schalk, M. Hannauer, and A. Braud, New roles for bacterial siderophores in metal transport and tolerance. Environ. Microbiol. **13,** 11 (2011). DOI: 10.1111/j.1462-2920.2011.02556.x

23. J. Sochor, O. Zitka, D. Hynek, E. Jilkova, L. Krejcova, L. Trnkova, and R. Kizek, Bio-Sensing of Cadmium (II) Ions Using *Staphylococcus aureus*. J. Sens. **11**, 11 (2011). DOI: 10.3390/s111110638

24. P. Kotchaplai, E. Khan, and A. S. Vangnai, Membrane Alterations in *Pseudomonas putida* F1 Exposed to Nanoscale Zerovalent Iron: Effects of Short-Term and Repetitive nZVI Exposure. Environ. Sci. Technol. **51**, 14 (2017). https://doi:10.1021/acs.est.7b00736

25. C. Lee, J. Y. Kim, W. I. Lee, K. L. Nelson, J.Yoon, and D. L. Sedlak, Bactericidal Effect of Zero-Valent Iron Nanoparticles on *Escherichia coli*. Environ. Sci. Technol. **42**, 13 (2008). https://doi:10.1021/es800408u

26. L. Chen, S. Jin, P. H. Fallgren, F. Liu, and P.J. Colberg, Passivation of zero-valent iron by denitrifying bacteria and the impact on trichloroethene reduction in groundwater. WST. **67**, 6 (2013). https://doi.org/10.2166/wst.2013.689

27. N. Desnoues, M. Lin, X. Guo, L. Ma, R. Carreño-Lopez, and C. Elmerich, Nitrogen fixation genetics and regulation in a *Pseudomonas stutzeri* strain associated with rice. J. Microbiol. **149**, 8 (2003). https://doi:10.1099/mic.0.26270-0

28. J. C. Gaby,and D. H. Buckley, A comprehensive aligned nifH gene database: a multipurpose tool for studies of nitrogen-fixing bacteria. Database: Database J. Biol. Databases Curation. (2014). https://doi.org/10.1093/database/bau001

29. J.-S. Chen, J. Toth, and M. Kasap, Nitrogen-fixation genes and nitrogenase activity in *Clostridium acetobutylicum* and *Clostridium beijerinckii*. JIMB. **27**, 5 (2001). https://doi:10.1038/sj.jim.7000083

30. R. Fani, R. Gallo, and P. Liò, Molecular Evolution of Nitrogen Fixation: The Evolutionary History of the *nif*D, *nif*K, *nif*E, and *nif*N Genes. J. Mol. Evol. **51**, 1 (2000). DOI: 10.1007/s002390010061

A comprehensive analysis of palm oil mill effluent as substrate biogas: A bibliometric approach

Rita Purwasih[1,2], *Sutaryo* Sutaryo[1], *Agung* Purnomoadi[1], and *Endang* Purbowati[1]

[1]Department of Animal Sciences, Faculty of Animal and Agricultural Sciences, Diponegoro University, 50275, Semarang, Indonesia
[2]Department of Agroindustry, Subang State Polytechnic, 41285, Subang, Indonesia

Abstract. The evolution, trend, and future development prospects of research on using Palm Oil Mill Effluent (POME) as a biogas substrate were observed. Results exhibited an average POME-focusing publication growth level of 21.74% per year. Chan Y. J. was the most relevant author with 26 publications, and Universiti Teknologi Malaysia was recorded as the most prominent affiliation with 104 publications. The most relevant source was AIP Conference Proceedings (31 articles) in 2005-2024. Malaysia and Indonesia were the two top countries in scientific collaboration and production related to this topic. The thematic map indicated evolution in various zones, with evolving research trends, especially in the fields of co-digestion, bioenergy, food waste, methane production, and circular economy. Results also confirmed the significant potencies of POME as a renewable energy source and in waste conversion into energy through biogas, with production optimization remaining open for further research.

1 INTRODUCTION

Palm oil industries are receiving much spotlight associated with environmental issues and impacts they are suspected of, covering deforestation, habitat degradation, and greenhouse gas release. On the other hand, global demand for palm oil is reportedly building up [1]. The global palm oil production in 2023-2024 reached 77.28 million metric tons, which increased by 0.94%, or 76.56 million metric tons in 2022-2023 [2]. [3] and [4] substantiate that palm oil processing industries with a production of 100 tons of fresh fruit bunches with an addition of 30 tons of water can produce 22.5 tons of crude palm oil (CPO) and by-products of empty fruit bunches (21.5 tons), shells (5.4 tons), fibers (13.1 tons), and palm oil mill effluent (POME) (62.5 tons). A concrete measure is thus entailed to combat or minimize any detrimental impacts to the environment.

Palm oil mill effluent (POME) is the by-product with the highest production. As it also contains high organic materials and comes with abundant material availability, many researchers prefer it as the staple material of new renewable energy sources through anaerobic digestion (AD). The preference is also a solution to ecological damage, including water pollution and aquatic ecosystem disruptions, which will break out once POME is disposed of to the environment directly [5][6]. Anaerobic digestion (AD) is a method considered effective for processing POME due to its strengths, encompassing short-time demand, high biodegradability, and yielded biogas as alternative energy through converting organic material compounds contained in POME [7][8][4]. Biogas yielded is useful in generating electricity and heat and can be upgraded to biomethane [9][10].

Now we can find ample research on using POME as a biogas substrate, but to the extent of our knowledge, there is an absence of bibliometric reviews to amass information from studies having been carried out. A bibliometric analysis is a scientific approach using a literature review conducted qualitatively and quantitatively to apprehend previous research findings [11]. A bibliometric analysis under the theme of using POME as a biogas substrate is therefore crucial to identify the trend of related research and the comprehensive conception of what it becomes in the future [12][13]. This research, accordingly, focuses on scrutinizing the trend, gap, and general description of the characteristics and strategies for the future development of research on using POME as a biogas substrate.

2 MATERIALS AND METHODS

A Scopus database with the keyword "palm AND oil AND mill AND effluent AND biogas" was invoked, resulting in 501 documents. The applied exclusion criteria included limited year (2005-2025) (12 items excluded) and document types (four items excluded). Document types used were article, book chapter, conference paper, conference review, and review, resulting in 485 documents as the database. The database was exported into a CSV-type file [14] and was analyzed using the R-Studio Bibliometrix software [15][16].

3 RESULT AND DISCUSSION

3.1 Main information about data

The annual growth level of publications under the selected research theme was 21.74% on average, classified as high ($R2 = 0.8322$). The document production trend in 2005-2009 was low, allegedly because the research topic still gained insubstantial popularity due to its novelty. Still, the trend exhibited an increase in 2010-2016 and reached the peak of publication in 2017-2024 with 54 documents in 2019.

Table 1. Main information about database from bibliometrik

Description	**Results**
Main information about data	
Timespan	2005:2024
Sources (Journals, Books, etc)	209
Documents	485
Annual Growth Rate %	21.74
Document Average Age	5.7
Average citations per documents	21.67
References	17748
Document contents	
Keywords Plus (ID)	2446
Author's Keywords (DE)	977
Authors	2109
Authors	1311
Authors of single-authored documents	7

Description	**Results**
Authors collaboration	
Single-authored documents	10
Co-Authors per Documents	4.67
International co-authorships %	31.34
Documents types	
Article	313
Book chapter	22
Conference paper	117
Review	6
Short survey	27

From our content document analysis, there were 2446 keywords and 977 author keywords used by 1,311 authors. The research theme of using POME as a biogas substrate received considerable international attention, as testified by collaboration author analysis, that many authors engaged with international collaborations (31.34%), while the average co-authorship per document was 4.67 with ten single authors. Document types identified were articles, book chapters, conference papers, reviews, and short surveys at the numbers of 313, 22, 117, 6, and 27, respectively. Information elaborated on the overview was structural descriptions of research fields. According to [17], a bibliometric method allowed for displaying some main information using bibliographic data from a publication database to produce a structural conception of a research field.

3.2 Assessment of authors and sources

During the research timespan of 2005-2024, there are the top 5 most relevant authors, affiliations, corresponding author's countries, and publishers illustrated in Fig. 1 and 2 (A). Chan Y.J., Ahmad A., O-Thong S., Trisakti B., and Hashim H., are the top 5 most relevant authors to the research theme, each author has an h-index of 11, 13, 13, 8, and 9 (Fig. 1 (A)). The author uses the h-index or Hirsch index as a benchmark in developing his work according to his scientific field [19]. The top 5 most relevant affiliations are Universiti Teknologi Malaysia, Prince Of Songkla University, University Of Nottingham Malaysia, Universiti Putra Malaysia, and Universitas Sumatera Utara, which have published 104, 89, 81, 71, and 52 articles, respectively

The top five corresponding author's countries were Malaysia (190 articles), Indonesia (87 articles), Thailand (63 articles), Japan (11 articles), and Denmark (six articles). Of the five countries, there were single-country publications (SCPs) and multiple-country publications (MCPs), with the average ratio expressed in % MCP. The SCP to MCP ratio served as an indicator of international collaborations [18]. Malaysia published most articles at 190 documents (39.3%), 126 SCPs, 64 MCPs, and 33.7% MCP. The second country was Indonesia with 87 articles (17.9%), 62 SCPs, 25 MCPs, and 28.7% MCP. The two countries were also included in five countries' collaboration world map, i.e., Indonesia with Japan (22 freq.), Malaysia with Japan (18 freq.), Malaysia with China (11 freq.), Malaysia with India (11 freq.), and Malaysia with Indonesia (11 freq.). International collaboration was a fruitful means of sharing knowledge and experiences between laboratories, hence enhancing research quality standards by integrating the diverse skills of researchers [19][20].

Table 2. Descriptive Analysis: Top Five Most Relevant Authors, Affiliations, Corresponding Author's Countries, and Sources

Rank	**Descriptive Analysis**

	Authors (N Articles)	Affiliation (N Articles)	Corresponding Author's Countries (N Articles)	Relevant Sources (N Articles)
1st	Chan Y. J. (26)	Universiti Teknologi Malaysia (104)	Malaysia (190)	AIP Conference Proceedings (31)
2nd	Ahmad A. (24)	Prince of Songkla University (89)	Indonesia (87)	Chemical Engineering Transactions (20)
3rd	O-Thong S. (21)	University of Nottingham Malaysia (81)	Thailand (63)	IOP Conference Series: Earth and Environmental Science (20)
4th	Trisakti B. (18)	Universiti Putra Malaysia (71)	Japan (11)	IOP Conference Series: Materials Science and Engineering (19)
5th	Hashim H. (17)	Universitas Sumatera Utara (52)	Denmark (6)	Journal of Cleaner Production (18)

Our source analysis confirmed the top five publishers which published documents related to the research theme, namely AIP Conference Proceedings (31 articles), IOP Conference Series: Earth and Environmental Science (20 articles), IOP Conference Series: Materials Science and Engineering (20 articles), Journal of Cleaner Production (19 articles), and Chemical Engineering Transactions (18 articles) (Table 2). As argued by [22], publisher analysis was an essential component of bibliometric analysis as it could provide a comprehensive understanding of available publication platforms regarding a particular research subject.

3.3 Document Assessment with Bibliometrix

Results of the analysis of the top five most globally cited documents related to the research theme are listed in Table 3. The analysis was imperative to build a fundamental knowledge framework regarding a specific field of study, providing researchers and readers with a roadmap of progress related to the research theme under investigation [22].

[23], an article review, was the most globally cited document at 310 quotations (20.67/year). The paper proposed a holistic review of POME processing methods in Malaysia and stated that POME could serve as renewable energy, i.e., biogas, which was eco-friendly and cost-efficient, through an integrated system. And yet, some problems following included the willingness of industries concerned to shift from the existing technology they considered more affordable and easier to operate. In addition, it was undeniable that POME conversion into methane called for high capital and adaptive technology. Still, using POME as a biogas staple material could elevate the ET sector in Malaysia in the future.

Table 3. Top Five Most Globally Cited Documents

No.	Authors (Year), Publisher	DOI	Total Citations	TC Per Year
1	Wu *et al.* (2010), J. Environ Manage [23]	10.1016/j.jenvman.2010.02.008	310	20.67
2	Ji *et al.* (2013), Renewable Sustainable Energy Rev. [24]	10.1016/j.rser.2013.06.008	284	23.67
3	Kushairi *et al.* (2018), J. Oil Palm Res. [25]	10.21894/jopr.2018.0030	200	28.57

4	Zinatizadeh *et al.* (2006), Water Res. [30]	10.1016/j.watres.2006.07.005	197	10.37
5	Ozturk *et al.* (2017), Renewable Sustainable Energy Rev. [31]	10.1016/j.rser.2017.05.111	184	23.00

Furthermore, the top five most frequent words based on author keywords, referring to our WordClouds analysis results, were palm oil mill effluent (146 occurrences), biogas (117 occurrences), anaerobic digestion (81 occurrences), POME (51 occurrences), and biogas production (37 occurrences). [19] defined a keyword as a phrase representing a research aspect underlying literature, and keyword analysis enabled the mapping of hot research topics, hence helping detect research orientation and gaps.

3.4 Thematic map

The thematic map of author keywords clustered by their correlation and evolution levels of scientific domains is suggested in Figure 1. The map was divided into four zones by centrality and density, defining the significance and evolution of the research themes. The four zones were Zone 1: motor themes, Zone 2: niche themes, Zone 3: basic themes, and Zone 4: emerging or declining themes. Colored bubbles on author keywords signified clusters, while their sizes indicated keyword frequency in each cluster.

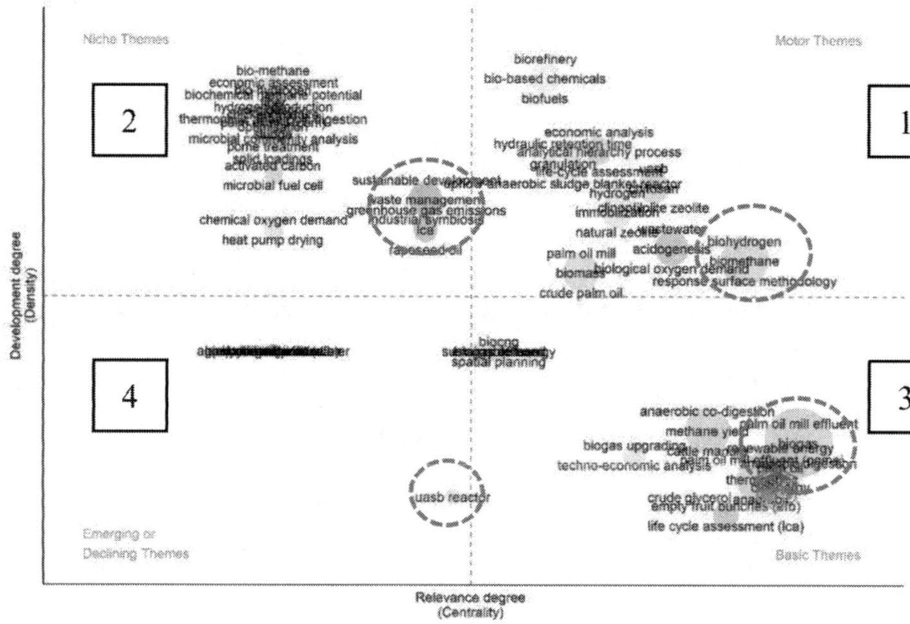

Fig. 1 Thematic map authors keyword by bibliometrix

For instance, palm oil mill, biomass, crude palm oil, climate change, greenhouse gas reduction, and life cycle impact assessment were included in Zone 1: motor themes, demonstrating that the keywords had high centrality and density, exhibiting their standing out and frequent uses. Meanwhile, palm oil mill effluent, biogas, anaerobic digestion, POME, biogas production, methane, co-digestion, empty fruit bunches, oil palm biomass, and microbial community were in Zone 3. They were identified with high centrality but low density, implying that albeit fundamental and substantial, they required further research. Keywords in Zone 3 could shift to Zone 1. UASB reactor was an example of keywords in

Zone 4, thus being defined as coming with low centrality and density, whereas hydrogen production, optimization, POME treatment, biogas plan, biological treatment, greenhouse gases, Monod model, and synthesis process were keywords in Zone 2. The latter group possessed high density but low centrality and was claimed as particular themes suitable to highly developing and isolated research topics [21].

3.5 Thematic evolution and trend topics

Figure 2 points out in chronology the thematic evolution of using POME as a biogas substrate by identifying clusters in the first timespan, i.e., 2005-2015, to the second one, i.e., 2016-2024. In the first timespan, 14 clusters were identified, yet during the second one, only seven were identified. It showed several phenomena. Firstly, only palm oil mill effluent existed in each timespan. Secondly, there was the merging of some clusters from the first timespan. For example, biodiesel, palm oil, and crude palm oil merged as renewable energy in the second timespan. Besides, several clusters of research topics either disappeared or appeared, signifying the time-to-time evolution of research on using POME in biogas. It corroborated the findings by [26], that the evolution of research was marked by merging, shifting, or the addition of topic clusters in each observation period. It was proof of the dynamic and evolutionary properties of research.

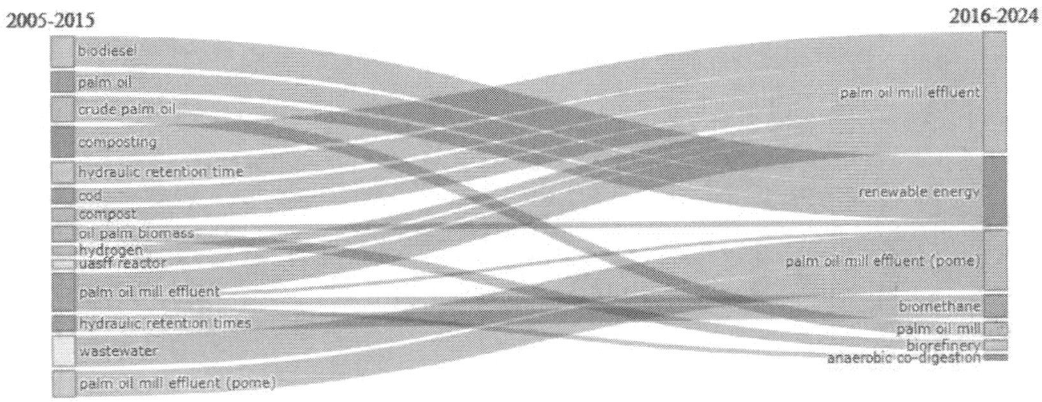

Fig. 2 Thematic evolution by bibliometrix

Thematic evolution analysis could be bolstered by trend topic analysis. [18] remarked that the latter analysis revealed interesting insights into research landscapes through the frequency and temporal distribution of many different research fields and highlighted time-to-time research evolution. Based on our analysis, the top five trend topics were palm oil mill effluent (146 frequency), highlighted mostly in 2015-2021, biogas (117 frequency), highlighted mostly in 2016-2022, anaerobic digestion (81 frequency), highlighted mostly in 2014-2021, POME (37 frequency), highlighted mostly in 2013-2022, and renewable energy (26 frequency), highlighted mostly in 2018-2021. Meanwhile, some trend topics in the latest years, which were 2021-2023, included co-digestion, bioenergy, food waste, methane yield and production, and circular economy. Looking into the results of our trend topic analysis, trend topics were of great benefit to studying the year-to-year evolution of research on using POME as a biogas substrate.

4 CONCLUSION

The bibliometric method could help identify the main themes and gaps of research on using POME as a biogas substrate. The average publication growth reached 21.74% per year. The main author was Chan Y. J., and Universiti Teknologi Malaysia was the most relevant affiliation. Malaysia and Indonesia were recorded as the main countries in research collaboration. The thematic map demonstrated research evolution in four zones, with the latest trend topics covering co-digestion, bioenergy, food waste, and circular economy. This research underlined the considerable potency of using POME in renewable energy and converting waste into energy by optimizing ever-developing biogas production.

CRediT authorship contribution statement

Rita Purwasih: Writing – original draft, investigation, formal analysis, preparation, methodology, data curation. **Sutaryo:** Supervision, funding acquisition. **Agung Purnomoadi:** Supervision, methodology. **Endang Purbowati:** Supervision, project administation.

Conflict of interest

The authors declare that they have no known competing financial interests or personal relationships that could have appeared to influence the work reported in this paper.

Data availability

Data will be made available on request.

Acknowledgment

The work has been carried out under the project (45839/S/09/2022), financed by the Indonesian Education Scholarship (BPI) who provided by Center for Higher Education Funding and Financing (BPPT) Ministry of Education, Culture, Research, and Technology (Kemendikbudristek), and Indonesia Endowment Fund for Education Agency (LPDP) Ministry of Finance (Kemenkeu) Republic of Indonesia.

References

[1] M.R. Zakaria, M. A. Ahmad Farid, H.S. Hafid, Y. Andou, and M. A. Hassan, Practical role of oil palm fronds in Malaysia's sustainable palm oil industry. Ind. Crops Prod. **222**, 1–19 (2024).

[2] USDA, 2024, Production-Palm Oil. Available at: https://fas.usda.gov/data/production/commodity/4243000 (Accessed: November 14, 2024).

[3] S. Supranto, A. Tawfiequrrahman, D. E. Yunanto, and I. Kurniawan, Oil palm empty fruit bunch fiber conversion to high refined cellulose using nitric acid and sodium hydroxide as the delignificating agents. AUN/SEEED-Net Reg. Conf. Chem. Eng. **1**, 1–15 (2014).

[4] I. Syaichurrozi, L. N. Sari, A. P. Apriantika, Jayanudin, N. Darsono, and D. S. Khaerudini, Effect of coagulation-sedimentation pretreatment on methane production from Indonesian palm oil mill effluent and kinetics. Bioresour. Technol. Reports **26**, 1–10 (2024).

[5] F. Alvionita, M. Faizal, L. N. Komariah, and M. Said, Biogas production from palm oil mill effluent with indigenous bacteria. Int. J. Adv. Sci. Eng. Inf. Technol. **9** (6), 2060–2066 (2019).

[6] A. Singkhala, C. Mamimin, A. Reungsang, and S. O-Thong, Enhancement of thermophilic biogas production from palm oil mill effluent by pH adjustment and effluent recycling. Processes **9** (5) (2021).

[7] M. Ahmadi-Pirlou, and T. Mesri Gundoshmian, The effect of substrate ratio and total solids on biogas production from anaerobic co-digestion of municipal solid waste and sewage sludge. J. Mater. Cycles Waste Manag. **23** (5), 1938–1946 (2021).

[8] T. M. Gundoshmian, and M. Ahmadi-Pirlou, Increasing biogas and methane yield by adding sewage sludge and zero-valent iron nanoparticles during the single-stage anaerobic digestion with municipal solid waste. Int. J. Energy Res. **46** (14), 20611–20623 (2022).

[9] J. Arjuna, T. B. Sitorus, H. Ambarita, and S. Abda, Performance and emissions of an engine fuelled by biogas of palm oil mill effluent. in IOP Conference Series: Materials Science and Engineering (2018).

[10] A. Fahrudin, D. F. Hakam, and W. Prawiraatmadja, Utilization of biomethane to support energy security in indonesia, riau province: A case study. Eur. J. Bus. Manag. Res. **9** (2), 99–106 (2024).

[11] M. Aria, and C. Cuccurullo, Bibliometrix: An R-tool for comprehensive science mapping analysis. J. Informetr. **11** (4), 959–975 (2017).

[12] A. B. Nasrin, S. K. Loh, M. A. Sukiran, N. A. Bukhari, and A. A. Aziz, Technical assessment and flue gases emission monitoring of an oil palm biomass–biogas cofired boiler. Environ. Prog. Sustain. Energy **38** (5) (2019).

[13] Heriyanti, A. L. Cahyani, A. Pertiwiningrum, I. M. Hidayatullah, and M. Gozan, Simulating biogas production process from palm oil mill effluent for power generation. E3S Web Conf. **503,** 1–8 (2024).

[14] D. Moher, A. Liberati, J. Tetzlaff, D.G. Altman, G. Antes, D. Atkins, V. Barbour, N. Barrowman, J.A. Berlin, J. Clark, M. Clarke, D. Cook, R. D'Amico, J.J. Deeks, P. J. Devereaux, K. Dickersin, M. Egger, E. Ernst, P.C. Gøtzsche, J. Grimshaw, G. Guyatt, J. Higgins, J.P.A. Ioannidis, J. Kleijnen, T. Lang, N. Magrini, D. McNamee, L. Moja, C. Mulrow, M. Napoli, A. Oxman, B. Pham, D. Rennie, M., Sampson, K.F. Schulz, P. G. Shekelle, D. Tovey, and P. Tugwell, Preferred reporting items for systematic reviews and meta-analyses: The PRISMA statement. PLoS Med. **6** (7), 1–6 (2009).

[15] H. S. Kusuma, D. E. Christa Jaya, N. Illiyanasafa, K. L. Ikawati, E. Kurniasari, H. Darmokoesoemo, and A. N. Amenaghawon, A critical review and bibliometric analysis of methylene blue adsorption using leaves. Chemosphere **356** (1-20) (2024).

[16] N. J. Van Eck, and L. Waltman, VOSviewer Manual, Man. VOSviewer version 1.6.20 Softw. Doc. 1–55 (2023).

[17] I. Zupic, and T. Čater, Bibliometric methods in management and organization. Organ. Res. Methods **18** (3), 429–472 (2015).

[18] S. I. Abdelwahab, M. M. E. Taha, A. A. Jerah, A. Farasani, S.M. Abdullah, I. A. Aljahdali, O. Oraibi, B. Oraibi, H. A. Alfaifi, A. H. Alzahrani, and Y. O. H. Babiker, Insights into frankincense and myrrh research: A comprehensive analytical study of patterns and perspectives. Heliyon **10** (19), 1–19 (2024).

[19] M. O. Jimoh, K. Okaiyeto, O.O. Oguntibeju, and C. P. Laubscher, A Systematic review on amaranthus-related research. Horticulturae **8** (3) (2022).

[20] F. G. Montoya, A. Alcayde, R. Baños, and F. Manzano-Agugliaro, A fast method for identifying worldwide scientific collaborations using the Scopus database. Telemat. Informatics **35** (1), 168–185 (2018).

[21] M. Akın, E. Bartkiene, F. Özogul, S.P. Eyduran, M. Trif, J. M. Lorenzo, and J. M. Rocha, Conversion of organic wastes into biofuel by microorganisms: A bibliometric review. Clean. Circ. Bioeconomy **6**, 1–11 (2023).

[22] M. H. Mohamed Noor, N. Ngadi, and N. H. Ab Hamid, Bibliometric insights into palm oil mill effluent treatment by coagulation-flocculation: Research trends and future directions. Ind. Crops Prod. **222**, 1–15 (2024).

[23] T. Y. Wu, A.W. Mohammad, J. M. Jahim, and N. Anuar, Pollution control technologies for the treatment of palm oil mill effluent (POME) through end-of-pipe processes. J. Environ. Manage. **91** (7), 1467–1490 (2010).

[24] C. M. Ji, P. P. Eong, T. B. Ti, C. E. Seng, and C. K. Ling, Biogas from palm oil mill effluent (POME): opportunities and challenges from Malaysia's perspective. Renew. Sustain. Energy Rev. **26**, 717–726 (2013).

[25] A. Kushairi, S. K. Loh, I. Azman, E. Hishamuddin, M. Ong-Abdullah, Z. B. M. N., Izuddin, G. Razmah, S. Sundram, and G. K. A. Parveez, Oil palm economic performance in Malaysia and r&d progress in 2017. J. Oil Palm Res. **30** (2), 163–195 (2018).

[26] H. D. H. Kariri, and A. Almubaddel, From theory to practice: revealing the real-world impact of cognitive behavioral therapy in psychological disorders through a dynamic bibliometric and survey study. Heliyon **10** (18), 1–19 (2024).

[27] R. Rodríguez-Soler, J. Uribe-Toril, and J. De Pablo Valenciano, Worldwide trends in the scientific production on rural depopulation, a bibliometric analysis using Bibliometrix R-tool. Land use policy **97**, 1–20 (2020).

[28] Z. K. Liew, Y. J. Chan, Z. T. Ho, Y. H. Yip, M. C. Teng, A. I. T. Ameer Abbas bin, S. Chong, P. L. Show, and C. L. Chew, Biogas production enhancement by co-digestion of empty fruit bunch (EFB) with palm oil mill effluent (POME): Performance and kinetic evaluation. Renew. Energy **179**, 766–777 (2021).

[29] M. K. Lam, and K. T. Lee, Renewable and sustainable bioenergies production from palm oil mill effluent (POME): Win-win strategies toward better environmental protection. Biotechnol. Adv. **29** (1), 124–141 (2011).

[30] A. A. L. Zinatizadeh, A. R. Mohamed, A. Z. Abdullah, M. D. Mashitah, M. Hasnain Isa, and G. D. Najafpour, Process modeling and analysis of palm oil mill effluent treatment in an up-flow anaerobic sludge fixed film bioreactor using response surface methodology (RSM). Water Res. **40** (17), 3193–3208 (2006).

[31] M. Ozturk, N. Saba, V. Altay, R. Iqbal, K. R. Hakeem, M. Jawaid, and F. H. Ibrahim, Biomass and bioenergy: An overview of the development potential in Turkey and Malaysia. Renew. Sustain. Energy Rev. **79,** 1285–1302 (2017).

Barriers to the Adoption of Biogas Technology in Smallholder Dairy Farming in Indonesia: A Case Study in Semarang Regency, Central Java

Bambang Haryanto[1,2], *Sutaryo* Sutaryo[2*], *Agung* Purnomoadi[2], *Endang* Purbowati[2], *Farhan Akmala Putra*[1,3], and *Chanifah* Chanifah[4]

[1]Research Center for Sustainable Production System and Life Cycle Assessment, National Research and Innovation Agency, Banten, 15314, Indonesia
[2]Department of Animal Science, Faculty of Animal and Agricultural Sciences, Universitas Diponegoro, Semarang, 50275, Indonesia
[3]Magister Agribisnis, Faculty of Animal and Agricultural Sciences, Universitas Diponegoro, Semarang, 50275, Indonesia
[4] Research Center for Behavioral and Circular Economics, National Research and Innovation Agency, Jakarta, 12710, Indonesia

Abstract. This study aims to identify and analyze the barriers faced by dairy farmers in adopting biogas technology in Manggihan Village, Semarang Regency. This village was selected as a case study due to its significant trend of increasing biogas installations. Farmers in Manggihan Village have a good understanding of biogas potential as an alternative energy source. They are aware of the benefits of biogas in reducing operational costs, preserving the environment, and obtaining an affordable energy source. However, the adoption of biogas technology still faces several challenges, particularly the high initial investment cost. The increased adoption of biogas in Manggihan Village is influenced by several factors, such as government programs, rising conventional energy prices, and growing awareness of environmental importance. Based on the research findings, it can be concluded that with appropriate support, particularly from the government, the potential of biogas as an alternative energy source in the dairy farming sector is very promising. In addition to reducing biogas installation costs, one strategic measure that can be undertaken is reallocating a portion of LPG subsidies to fund biogas installations. This will provide incentives for farmers to switch to biogas and reduce reliance on fossil fuels.

1 Introduction

Energy is one of the basic human needs that plays a crucial role in supporting various life activities. Advances in technology, coupled with population growth, have created a new era of life that drives increased energy demand (Kandpal and Singh, 2022). Biogas holds potential as an alternative energy source for farmers. It is produced through the anaerobic

* Corresponding author: sutaryoundip@yahoo.com

digestion of organic materials, such as human and animal waste, organic waste, household waste, and other biodegradable materials (Sahara, 2024). In addition to serving as an energy source, biogas offers several benefits, including reducing greenhouse gas emissions, improving environmental quality, and producing organic fertilizer (Kasap, 2012). Anaerobic digestion is a sustainable method for waste management, providing environmental protection and economic savings by converting organic matter into biogas. This process also recovers resources and energy, reduces pollution from agricultural and industrial operations, and simultaneously offsets fossil fuel usage in operations (Chen, et al., 2008).

Approximately 90% of dairy farming in Indonesia is conducted by smallholder farmers, each owning 3 to 6 dairy cows (Susilorini, et al., 2022). These farms are typically small-scale and rely on simple technology. Smallholder dairy farming is of interest to agricultural policymakers in Indonesia, inspired by its success in South Asian countries, despite facing several major challenges (Remenyi, 2006). The increasing population of dairy cows, predominantly managed in smallholder farms, has intensified the negative impacts of poor manure management, leading to substantial amounts of untreated livestock waste (al Zahra, et al., 2020). It is estimated that 84% of dairy manure is disposed of without any processing, causing environmental pollution (De Vries and Wouters, 2017). The dairy cow population in Indonesia is projected to grow by 16.01% from 2022 (507,075 cows) to 588,280 cows in 2026, with milk production increasing by 15.22% from 824,273 tons in 2022 to 949,733 tons (Pusdatin, 2022).

The potential for biogas development is immense, given the large population of dairy cows. Livestock waste represents an abundant raw material for biogas production, which can serve as an alternative energy source (Sahara, 2024). However, the implementation of biogas technology in smallholder farms in Indonesia remains suboptimal due to resource and knowledge limitations (Sahara, 2024). The adoption of biogas technology in smallholder farming is further hindered by challenges such as high investment costs and perceived complexity (Dyah and Sriharti, 2019).

This study aims to identify and analyze the barriers faced by smallholder dairy farmers in Manggihan Village, Getasan District, Semarang Regency, in adopting biogas technology. The village was selected because it exhibits a growing number of biogas installations and is considered more successful compared to other villages. The findings of this study are expected to represent the challenges faced by smallholder dairy farmers in other regions of Indonesia. Additionally, the study aims to formulate policy recommendations and strategies to address these challenges and enhance biogas utilization in smallholder farming systems.

2 Metodologi

The study was conducted in Manggihan Village, Getasan District, Semarang Regency, Central Java Province, during November and December 2024. This research utilized both primary and secondary data. Primary data collection was carried out through surveys using structured questionnaires, while in-depth interviews were conducted to complement the findings. A total of 32 smallholder dairy farmers from Manggihan Village participated in the study.

Purposive sampling was employed to gather primary data regarding farmers' characteristics and their perceptions of the use and adoption of biogas. Secondary data were also included to support the discussion. These secondary data were obtained from various sources, including the Manggihan Village profile, websites, and various research publications. These sources provided information on (1)The geographical conditions of Manggihan Village, (2) Population data and the number of dairy farmers, (3) The population of dairy cattle, and (3) The number of biogas installations.

Analysis of Farmers' perception terhadap Biogas
Farmers' perceptions of biogas were analyzed based on multiple attributes, encompassing both attributes of biogas usage as a product and farmers' willingness to adopt biogas. The multi-attributes of biogas usage as a product include: (i) the use of biogas as a substitute for LPG, (ii) the ease of biogas production technology, (iii) consumer availability (market share), (iv) the affordability of the product for consumers, and (v) the ease of marketing bioethanol products.
Meanwhile, the multi-attributes of farmers' willingness to adopt biogas include: (i) community acceptance of biogas technology, (ii) affordability of technology adoption costs, (iii) dissatisfaction with existing practices, (iv) input limitations, (v) land availability, (vi) sufficiency of technological information, and (vii) availability of additional workforce.
Perception data were initially qualitative ordinal data, which were then converted into quantitative data through a scoring (scaling) technique and subsequently averaged [Chikuta, 2014; Widodo, 2022]. Farmers' perceptions of the attributes of biogas usage and adoption levels were categorized into four scales, ranging from 1 to 4, with categories of very poor, poor, good, and very good, as shown in Equation 1 [Milkias, 2019].

$$Score = \frac{n_i.s_i}{N_i}$$

Where n_i is the number of respondents in column $_i$ $(_{i=1,2,3})$, s_i is the score of statement $_i$ $(_{i=1,2,3})$, and N_i is the number of respondents in row $_i$ $(_{i=1,2,3})$.
Subsequently, the performance score for the multi-attributes is determined within a 4-scale interval, as shown in Equation 2.
:

$$Interval\ scale = \frac{highest\ score\ -\ lowest\ score}{number\ of\ interval\ scale}$$

Attributes with higher scores are perceived as good, while those with lower scores are perceived as poor. The assessment of farmers' perception categories is based on the mean of attributes, which are mapped into four perception categories, as follows: 1)Very low perception with a mean score of 1.00–1.75, 2) Low perception with a mean score of 1.76–2.50, 3) High perception with a mean score of 2.51–3.25, and 4) Very high perception with a mean score of 3.26–4.00.
Farmers' views on these four products are illustrated using perceptual mapping through radar charts [Gigauri, 2019].

3 Results and Discussion

3.1 Farmers Caracteristics
Manggihan Village, located in Getasan District, Semarang Regency, is one of the villages in Semarang Regency. It is situated at coordinates 110°14'54.75" – 110°39'3" East Longitude and 7°3'57" – 7°30'0" South Latitude. The elevation of Manggihan Village is 800 meters above sea level (masl), with the lowest elevation in Seturun Hamlet and the highest in Sengon Hamlet (Manggihan, 2024). Manggihan Village consists of 599 households, 246 of which are dairy farmers. Among these dairy farmers, 62 households (or 25.20%) have installed biogas systems. Of all the dairy farmers, 85.16% (218) are smallholders, with an average ownership of 3.7 dairy cows per farmer. According to Matondang and Rusdiana, (2013), more than 90% of farmers in Indonesia are smallholder farmers characterized by relatively small-scale operations, typically ranging from 1 to 5 cows. The characteristics of smallholder farmers in Manggihan Village are presented in Table 1.

Tabel 1. Farmers Characteristics of Manggihan village.

No	Parameter's	
1	Farmer's age (years)	49.41 ± 13.29
2	Farming Experience (years)	25.97 ± 14.93
3	Education	
	Elementary school,	50.00%
	Junior high school	18.75%
	Senior high school	28.13%
	University	3.13%
4	Family income (USD)	155,22 ± 60,73
5	Primary Job	
	Farmer	50.00%
	Employee	15.38%
	Livestock farmer	23.08%
	Trader	11.54%
6	LPG requirement (kg/month)	9

Source: Research results, 2024

The average age of dairy farmers in Manggihan Village is slightly higher than that of farmers in Lembang, West Bandung Regency (44.86 years) (Sudrajat, et. al., 2022) and the overall average age in West Java, which is around 46 years (Akzar et al., 2022). The average age of farmers in Manggihan, reaching 49 years, indicates that their physical capacity tends to decline, which often results in suboptimal livestock maintenance (Sudrajat, et. al., 2022).

The experience of dairy farmers in Manggihan Village is similar with Fadillah, et al., (2023) reported that the average farming experience of dairy farmers in West Java is 19.1 years. Research by Satiti, et al., (2022) stated that farmers in Samiran Village, Boyolali Regency, predominantly have more than 15 years of farming experience.

The education level of farmers in Manggihan Village is dominated by primary education. This finding is consistent with research conducted in the same regency by Haloho et al., (2013), where 51.3% of farmers were primary school graduates. However, it differs from findings in other regions by Satiti, et al., (2022) and Nurdiyansah, et al., (2020), where farmers' education levels were dominated by high school graduates at 77.27% and 40%, respectively. Education and farming experience influence awareness and the adoption of technology (Takanjanji and Kaka, 2022). Farmers with longer experience and higher education levels tend to exhibit greater awareness and higher rates of technology adoption (Akzar, et al., 2022).

The primary occupation of most farmers in Manggihan Village is farming, while only 23.08% are primarily livestock farmers. This aligns with Satiti, et al., (2022) findings, which reported that 45.45% of farmers are primarily crop farmers, while 31.82% are primarily livestock farmers. Most small-scale dairy farming operations are managed as secondary businesses, leading to suboptimal attention to their management (Haloho et al., 2013).

This is reflected in the average income of farmers, which remains below the 2024 Semarang Regency Minimum Wage of USD 159.20 (Central Java Province, 2024). Low income has driven many young people in Manggihan Village to work in factories around Semarang and Salatiga. Parents play a crucial role in educating and raising awareness among the younger generation about the potential of dairy farming and the utilization of by-products to support family income (Firman et al., 2018). This is essential to ensure the sustainability of the livestock sector in Manggihan Village, so it does not rely solely on factory jobs in the Semarang and Salatiga areas.

3.2 Farmer's Perception on Biogas

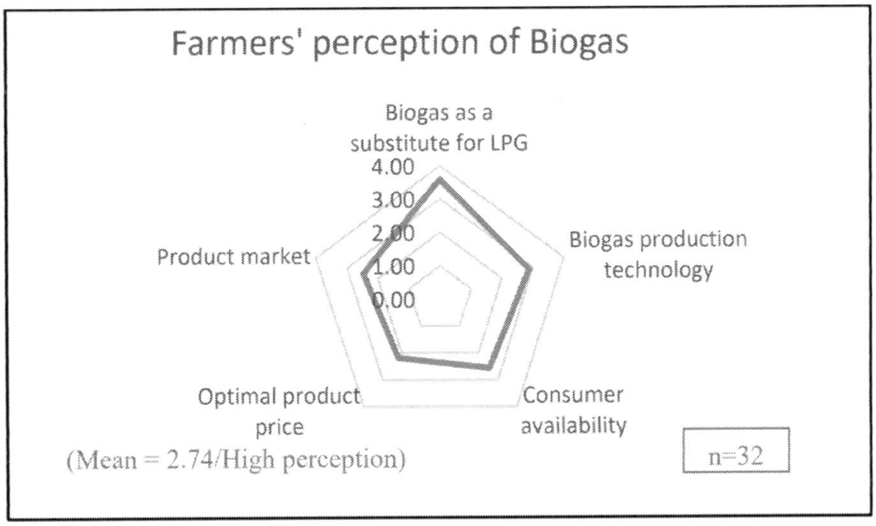

Source: Primary data 2024 (processed)

Note: scala/category:
a. 1,00-1,75 = Very low perception
b. 1,76-2,50 = low perception
c. 2,51-3,25 = High perception
d. 3,26-4,00 = Very high perception

Figure 1. Farmers' perceptions of biogas

Farmers have a high perception of biogas, primarily driven by the benefits they perceive from its use in daily life. One of the main factors contributing to this perception is the potential of biogas as a substitute for LPG. Farmers are consumers of subsidized LPG from the government, and there are certain moments when LPG gas becomes scarce (Kompas, 2023). In such situations, biogas significantly helps meet the energy needs of farmers. The use of biogas can save on expenses and provide long-term positive impacts on household economics (Savitri, et al., 2024). Circular economy principles in biogas can optimize the use of biomass resources (Situmeang, et al., 2022). Biogas is seen as a more cost-effective, environmentally friendly, and easily accessible alternative energy source for farmers, especially in rural areas. These factors contribute to the increasingly positive perception of biogas, both in terms of its attributes as an LPG substitute and its consumer availability.

The technology for biogas production is also an important factor that supports the positive perception. Farmers believe that the technology for producing biogas has become easier to implement, thanks to the available training and technical guidance. This perception differs from the research by Ariyanti, el at., (2024), which mentioned that beef cattle farmers in Karanganyar Regency perceived biogas technology as difficult to replicate. The positive perception in Manggihan Village is due to the active role of village officials and agricultural extension workers in socializing and educating farmers about biogas technology. The interaction between farmers and information sources, such as extension workers, plays a crucial role in the process of biogas technology transfer, helping farmers obtain sufficient information and knowledge to enhance their understanding of the technology (Putra. et al., 2019).

However, there are some factors that contribute to a lower perception of biogas among farmers, such as the price of biogas products and market access. These results align with the research by Arianti, et al., (2024), which found that farmers still consider biogas to be relatively expensive and difficult to market. Biogas technology is often perceived as requiring high initial investment and maintenance costs (Tigabu, et al., 2015), leading farmers to perceive the price of biogas as high and its marketing as challenging.

3.3 Farmers Perception of Biogas Adoption

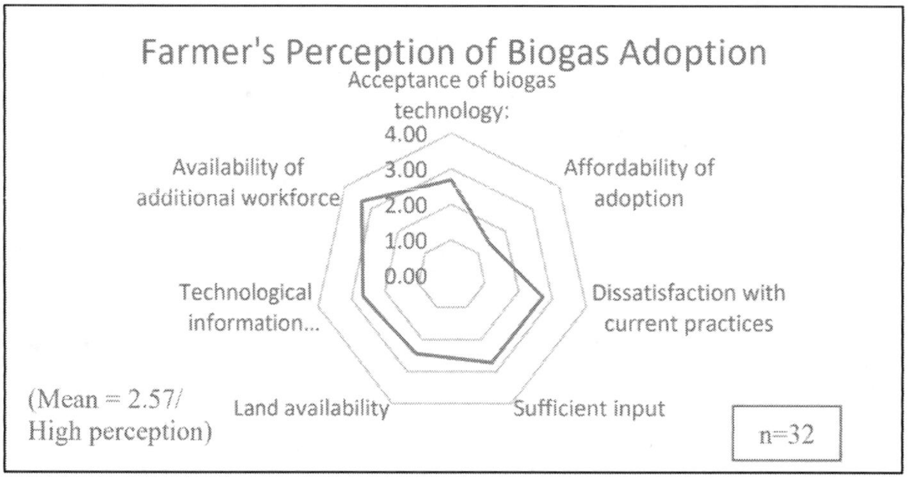

Source: Primary data 2024 (processed)

Note: scala/category:
a. 1,00-1,75 = Very low perception c. 2,51-3,25 = High perception
b. 1,76-2,50 = low perception d. 3,26-4,00 = Very high perception

Figure 2: Dairy Farmers' Perception of Biogas Adoption in Manggihan Village

Farmers have a high perception of biogas adoption. Factors such as acceptance of biogas technology, dissatisfaction with current practices, sufficient raw materials, ease of technological information, and availability of additional labor contribute significantly to this perception. A study in Paser Belengkong found a significant positive relationship between farmers' perceptions and the implementation of biogas technology (Tohri, et al., 2019). This perception is important because individuals' views on relative advantages and their experiences with biogas installations significantly contribute to the acceleration of technology adoption among farmers (Putra, et al., 2019). According to Ahmad, et al., (2023), the decision to adopt biogas in rural households is significantly influenced by relative advantages, trialability, and observability. Regarding the availability of labor, some previous biogas users were former firewood users, and the labor required for biogas operations is less than the labor required to gather firewood.

However, land availability contributes only slightly, while the affordability of adoption costs has a very low contribution to farmers' perception of biogas adoption. Smallholder households are characterized by low incomes with limited assets and resources, including small land (Purwantini, et al., 2021). Farmers in Manggihan typically keep cattle behind their homes, and some farmers live in densely populated areas with limited land, making it difficult to install biogas systems.

The upfront costs of biogas installations are substantial, which is a primary impediment to widespread adoption, especially among small-scale farmers (Diouf and Miezan, 2019). It should be noted that all biogas installations in Manggihan Village are provided by the government, and dairy farmers do not incur any costs for installation. There are several reasons why dairy farmers generally struggle to finance biogas installations independently:

A) High initial investment costs, according to Wahyudi (2017), the investment required to build a biogas digester is quite high, and some farmers believe it is more profitable to use the funds to increase livestock numbers or purchase land than to invest in building a biogas digester. B) Long payback period, the payback period for a biogas digester compared to LPG as a cooking energy source, calculated without subsidies and without commercial interest loans, shows that the investment in a biogas digester will be recovered in about 6 years (Atmaja, et al., 2014). C). Limited access to capital: Many farmers, especially small-scale farmers, have limited access to funding sources such as bank loans (Purwantini, et al., 2021). One potential solution is to create more affordable biogas installation packages. Currently, the biogas package installed in Manggihan Village costs about 1,000-1,200 USD. Farmers hope that the price of one biogas installation package can be reduced to around 300-400 USD. Currently, the Indonesian government provides a substantial amount of funding for energy subsidies, including the subsidy for LPG purchases. The LPG subsidy for 3 kg gas in Indonesia in 2024 is 4.915 million USD. The actual price of a 3 kg LPG gas is 2.62 USD, with a subsidy of 1.84 USD, allowing the community to purchase it for 0.78 USD (Antaranews, 2025). Farmers who do not use biogas typically consume an average of 3 cylinders of 3 kg LPG gas per month. If this subsidy were redirected to support biogas installation, it would only be needed once over several years, and the LPG subsidy for farmers in the coming years could be eliminated. There are currently cheap biogas installation alternatives with prices around 300 USD, which would be equivalent to the subsidy for 4 to 5 years.

Biogas has a positive impact on improving environmental quality and generating direct economic savings in rural areas (Atmaja, et al., 2014). Manggihan Village is one of the areas contributing to livestock waste runoff into Rawa Pening Lake. This lake, located in Semarang Regency, is the confluence of 14 tributaries, a volume of 48.15 million m3, and a sedimentation rate of 778.93 tons per year (Jatengprov, 2024), with revitalization costs from the Ministry of Public Works and Housing in 2024 amounting to 4.4 million USD (LKPP, 2023). It is time for efforts to stop sedimentation and the growth of water hyacinth upstream to begin. The costs of cleaning Rawa Pening Lake from water hyacinth and dredging sedimentation are very high. It would be more practical to redirect those funds for biogas development in upstream villages like Manggihan and the surrounding villages.

3.4 Biogas Adoption

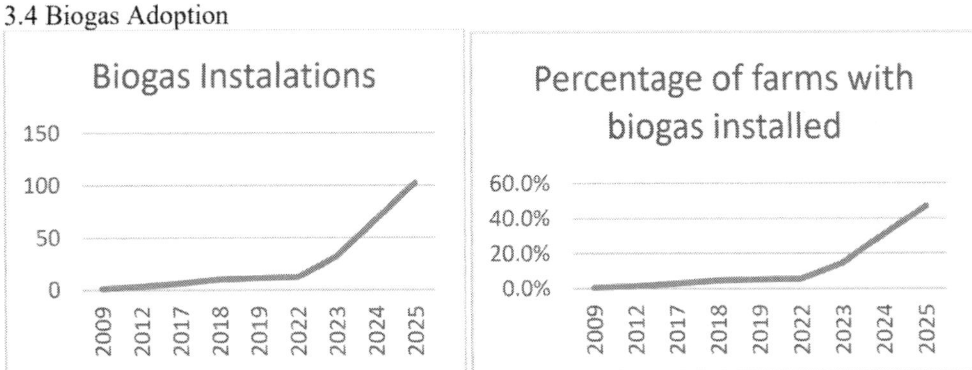

Figure 3: Trend of Biogas Installation Growth and Percentage of farms with biogas installed in Manggihan Village

Based on Figures 3, the adoption of biogas technology continues to increase. The growing number of biogas installations indicates that more dairy farmers in Manggihan Village are adopting this technology. This may be due to increased awareness among farmers about the benefits of biogas, such as cost savings, clean energy, and better waste management. Factors influencing biogas adoption among smallholder farmers in Indonesia include education,

female involvement, livestock ownership, income, availability of funds, and engagement with stakeholders (Putra, et al., 2019). The sharp increase in 2022 and 2024 suggests that there were factors driving the acceleration of biogas adoption during these periods.
Factors that may have contributed include: a. Increased biogas productivity, with the addition of a desulfurizer tool, making farmers more satisfied with the biogas equipment they have. b. Government programs. Biogas from agricultural waste can play a role in the energy transition, but it requires state incentives (Kabeyi and Olanrewaju, 2021). c. Rising conventional energy prices: The increasing cost of conventional energy such as LPG may have encouraged farmers to seek more affordable alternatives like biogas. d. Increased environmental awareness: Growing awareness of the importance of the environment and clean energy may also be a driving factor.
Potential for further growth: The continued upward trend through 2024 indicates that there is still significant potential for the growth of biogas installations in Manggihan Village. In 2025, it is expected that a proposal for the installation of 35 biogas units will be submitted. Currently, Manggihan Village has just introduced a new biogas system made of rubber, with an installation cost of 300 USD. The response from farmers is still unknown as the installation is limited to 5 farmers in December 2024. Hopefully, with the reduction in installation costs, this will encourage more farmers to adopt biogas technology in Manggihan and other regions in Indonesia.

4 Conclusion

Small Holder Farmers in Manggihan Village, Semarang District, has high perception on biogas as a product and willingnes to biogas adoption, reflecting a relatively high level of acceptance. However, some farmers face challenges due to limited land availability. The primary barrier to adopting biogas technology is the inability of farmers to afford the necessary biogas equipment. The government should recognize biogas as an opportunity to promote renewable energy while achieve long-term savings on 3 kg LPG subsidies. In addition to efforts to reduce biogas installation costs, government subsidies could provide significant support to small-scale farmers.

Acknowledgements

The author would like to Express their gratitude to the Indonesian Endowment Fund for Education (Lembaga Pengelola Dana Pendidikan, LPDP) for supporting the writing process of this article.

References

Antaranews. 2015. Realisasi subsidi BBM dan LPG 2024. https://www.antaranews.com/infografik/4571214/realisasi-subsidi-bbm-dan-lpg-2024 Accessed 27 Desember 2024.

A. D. Tigabu, F. Berkhout, P. van Beukering,. Technology innovation systems and technology diffusion: Adoption of bio-digestion in an emerging innovation system in Rwanda, Technological Forecasting and Social Change, Vol90, Part A, (2015)

A. Fadillah, B. H. P. van den Borne, O. N. Poetri, H. Hogeveen, W Umberger, J. Hetherington, and Y. H. Schukken.. Smallholder milk-quality awareness in Indonesian dairy farms. J. Dairy Sci. 106:7965–7973 (2023)A. Firman, L. Budimulati, M. Paturochman, M. Munandar. Succession models on smallholder dairy farms in Indonesia. Livestock Research for Rural Development 30 (10) (2018).

A. Kasap, R. Aktas, E. Dulger, Economic and Environmental Impacts of Biogas. A. Kasap, R. Aktas, E. Dulger, Economic and Environmental Impacts of Biogas. A. Kasap, R. Aktas, E. Dulger, Economic and Environmental Impacts of Biogas.

A. Sudrajat, L. Amin, R. F. Christi, R. Sambodo, F. Ismail. Profil Peternak Sapi Perah di Lembang Kabupaten Bandung Barat. Jurnal Sumber Daya Hewan Vol. 3 No. 2 (2022).

A. R. S. Putra, T. G. Czekaj, M. Lund, Study of the biogas technology adoption as a livestock waste management among smallholder farmers in Indonesia. *IOP Conf. Ser.: Earth Environ. Sci.* 260 012070, (2019)

B. Diouf, Miezan E. The biogas initiative in developing countries, from technical potential to failure: The case study of Senegal. Renewable and Sustainable Energy Reviews, 101, pp. 248 – 254 (2019).

C. Meidiana, A. Hidayah, W. P. Wijayanti, Factors Influencing the Biogas Acceptance in Rural Area. IOP Conf. Ser.: Earth Environ. Sci. 520 012006, (2020)

D. F. Shams, M. Qasim, S. Khalid, K. Shah, Impact of Biogas on Sustainable Livelihood in Rural Areas A Case Study of Swat, Pakistan. (2014)

D. Sahara, Livestock waste as alternative energy for rural households: A review. IOP Conf. Ser.: Earth Environ. Sci. 1292 012002 (2024)

E Satiti, S Andarwati, T A Kusumastuti, The Role of Woman in Livestock Farmers' Groups in Samiran, Boyolali, Central Java. Jurnal Kawistara, The Journal of Social Sciences and Humanities. Vol. 12, No. 1, 30 April 2022: 79-98. (2022).

F. D. Arianti, J. Triastono, M. D. Pertiwi, A. Prabowo, T. Prasetyo, Chanifah, B. Haryanto, Megawati, W. Astuti, I. N. Djarot, A D Santoso, S. P. Wijayanti. Renewable energy potential of rice straw and paunch manure as bioethanol feedstocks in Central Java, Indonesia. Case Studies in Chemical and Environmental Engineering,Volume 9, (2024)

H. R. Matondang, S.Rusdiana. Langkah-Langkah Strategis dalam Mencapai Swasembada Daging Sapi/Kerbau 2014. Pusat Penelitian dan Pengembangan Peternakan Depertemen Pertanian. Bogor (2013)

I. Nurdiyansah, D. Suherman, H. D. Putranto. Hubungan Karakteristik Peternak dengan Skala Kepemilikan Sapi Perah di Kecamatan Kabawetan Kabupaten Kepahiang. Bulletin of Tropical Animal Science. 1.(2): 64-72, (2020)

Jatengprov.go.id. World Water Forum" Bali 2024, Pj Gubernur Jateng Beberkan Pengelolaan Danau Rawa Pening yang Berkelanjutan. Accessed 27 Desember 2024. (2024).

Jateng Prov.go.id. https://jatengprov.go.id/publik/umk-jateng-2024-diumumkan-kota-semarang-tertinggi/ Accessed 27 Desember 2024.

J. V. Remenyi, Issues in Smallholder Tropical Dairying. Bulletin of Indonesian Economic Studies Volume 22, 1986 - Issue 1. https://doi.org/10.1080/00074918612331334764 (2006)

J. Wahyudi,. The Determinant Factors of Biogas Technology Adoption in Cattle Farming: Evidences from Pati, Indonesia. Int. Journal of Renewable Energy Development 6 (3) 2017: 235-240 (2017)

Kompas. Elpiji 3 kg langka di beberapa daerah ini akar penyebabnya https://money.kompas.com/read/2023/08/06/111236626/elpiji-3-kg-langka-di-beberapa-daerah-ini-akar-penyebabnya#google_vignette Accessed 27 Desember 2024. (2023).

K. Takanjanji, A. Kaka. Pengaruh Karakteristik Peternak Terhadap Adopsi Teknologiinseminasi Buatan Pada Ternak Babi Di Kelurahan Matawai Kecamatan Kota Waingapu. Jurnal Peternakan Sabana 1 (2) (2022)

Manggihan, 2024. Manggihan Village Provile. (2012).

M. de Vries, B. Wouters, Characteristics of small-scale dairy farms in Lembang, West-Java. (Livestock Research report; No. 1076). Wageningen Livestock Research. (2017).

M. J. Kabeyi, O. A. Olanrewaju, Development of a Biogas Plant with Electricity Generation, Heating and Fertilizer Recovery Systems. (2021).

M. Ahmad, I. Khan, M. Q. S. Khan, G. Jabeen, H. S. Jabeen, C. Işık. (2023) Households' perception-based factors influencing biogas adoption: Innovation diffusion framework. Energy Vol. 263, Part E, 15 January

M. Tohri, J. R. Manulang, Mursidah. Perception of the Breeders of Beef Cattle on the Utilization of Biogas Technology in Paser Belengkong District. Jurnal Peternakan Lingkungan Tropis, VOL. 2 N0. 1 (2019)

R. Akzar, W. Umberger, A. Peralta. Understanding heterogeneity in technology adoption among Indonesian smallholder dairy farmers. Agribusiness. 2023;39:347–370. (2022)

R. D. Haloho, S. I. Santoso, S. Marzuki. (2013) Efisiensi Usaha Peternakan Sapi Perah Di Kabupaten Semarang. Agromedia, Volume 31, Nomor 2 September 2013.

R. Kandpal, R. Singh, Renewable Energy Sources – A Review. *ECS Trans.* 107 8133. (2022)

R Situmeang, J. Mazancová, H. Roubík, Technological, Economic, Social and Environmental Barriers to Adoption of Small-Scale Biogas Plants: Case of Indonesia. *Energies*, *15*(14), 5105. (2022).

S. Atmaja, P. Rosyidi, T. Bole-Rentel, S. B. Lesmana, J. Ikhsan. Lessons Learnt from the Energy Needs Assessment Carried out for the Biogas Program for Rural Development in Yogyakarta, Indonesia. Procedia Environmental Sciences Volume 20, (2014)

S. Dyah, and Sriharti. Biogas development: Dissemination and barriers. IOP Conf. Ser.: Earth Environ. Sci. 277 012018. (2019)

T. B. Purwantini, H. P. Saliem, E. Ariningsih, Erwidodo, I. S. Anugrah, E. Suryani, A. R. Irawan and J. B. Hetherington. The performance of smallholder dairy farms in West Java. OP Conf. Ser.: Earth Environ. Sci. 892 012098 (2021)

T. E. Susilorini, P. Surjowardojo, R. D. Wahyuni, Suyadi, Good dairy farming practices (GDFP) implementation on smallholder dairy farmers in East Java, Indonesia. Jurnal Ilmu-Ilmu Peternakan 32(1): 118 – 129 (2022).

U. R. Savitri, A. Purnomo, Y. A. Prasetyo, G. Rifai, M. Prasetya. Dampak Positif Pemanfaatan Kotoran Sapi Sebagai Biogas Dalam Perekonomian Rumah Tangga. Kridacendekia Vol 3 No 1 April 2024-Juli2024. (2024)

W. Zahra, E. Corina, van Middelaar, J. M. Imke, de Boer, J. Simon Oosting. Redicting nutrient excretion from dairy cows on smallholder farms in Indonesia using readily available farm data. Asian-Australas J Anim Sci. Vol. 33, No. 12:2039-2049 December 2020 https://doi.org/10.5713/ajas.20.0089 (2020)

Y Chen, Cheng JJ, Creamer KS Inhibition of anaerobic digestion process: a review. Bioresour Technol 99(10):4044–4064. (2008)

Rapid detection of adulteration in red dates (*Ziziphus jujuba* mill.) powder by Fourier-transform infrared spectroscopy combined with chemometrics analysis

Tai Wern Chze Crystal[1], *Khye Er* Loh[1*], *Asmiyenti Djaliasrin* Djalil[2], *Wiranti* Sri Rahayu[2], *Ghim Hock* Ong[3]

[1]Department of Bioscience, Faculty of Applied Sciences, Tunku Abdul Rahman University of Management and Technology, Jalan Genting Kelang, Setapak, 53300 Kuala Lumpur, Malaysia
[2]Fakultas Farmasi, Universitas Muhammadiyah Purwokerto, Jl. KH. Ahmad Dahlan, Dusun III, Dukuhwaluh, Kec. Kembaran, Kabupaten Banyumas, Jawa Tengah 53182, Indonesia
[3]Faculty of Health and Life Sciences, INTI International University, Persiaran Perdana BBN, Putra Nilai, 71800 Nilai, Negeri Sembilan, Malaysia

Abstract. Food adulteration presents a major challenge in the food industry, impacting product quality and consumer trust. This study investigates the use of Fourier Transform Infrared (FTIR) spectroscopy combined with chemometric methods, Principal Component Analysis (PCA) and Partial Least Squares-Discriminant Analysis (PLS-DA), for detecting cornstarch and powdered sugar adulterants in *Ziziphus jujuba* Mill. fruit powder. The PCA model effectively separated pure and adulterated jujube powders, clustering samples by adulterant type and concentration. Brands A, B, and E of the retail jujube powder samples were identified as likely adulterated with corn starch, while Brands C and D showed minimal adulteration. PLS-DA further differentiated adulteration levels, suggesting high corn starch adulteration levels in Brand A, B, and E samples. A Partial Least Squares Regression (PLSR) model yielded high accuracy (R^2 = 0.9859) and precision, with low error rates (RMSEE = 5.50% and RMSECV = 9.16%), supporting the model's robustness in adulteration prediction. Our findings highlight FTIR spectroscopy coupled with chemometric modelling as an efficient, reliable tool for rapid adulteration detection in jujube powder. This approach offers significant potential for ensuring product authenticity and quality control in functional food products.

1 Introduction

Food fraud, encompassing a wide range of deceptive practices within the food supply chain, incurs substantial economic losses, estimated at $30-40 billion annually [1]. One of the most common types of food fraud is food adulteration. It is the act of intentionally degrading food quality by adding or substituting food ingredients with undisclosed replacements, or by

* Corresponding author: lohke@tarc.edu.my

removing an essential ingredient. This pervasive issue manifests in diverse forms, as evidenced by the adulteration of honey with sugar [2], the substitution of cheaper oils for olive oil [3], and illegal dye and chemical use in spices [4]. Such practices intentionally incorporate low-cost fillers like cornstarch and powdered sugar, diminishing the product's quality, safety, and economic worth. The potential health risks associated with food fraud necessitate a paradigm shift. Global estimates suggest that approximately 22% of the food supply is tainted each year, resulting in health issues for nearly 57% of consumers [5].

Ziziphus jujuba Mill., commonly referred to as red date, Chinese date, or jujube, has been held in high esteem for centuries due to its medicinal and nutritional properties [6]. Jujube is utilised in various culinary applications and beverages as a natural sweetener as it lends its unique flavours to recipes. Adulteration of *Ziziphus jujuba* Mill. fruit powder represents a pressing concern that erodes consumer trust and jeopardises public health. Detection and prevention of adulteration in jujube fruit powder is paramount to maintaining the integrity of the industry and safeguarding consumer well-being. Traditional methods for identifying adulteration can be time-consuming, costly, and may lack the requisite sensitivity and specificity [7]. This research aims to answer whether FTIR spectroscopy, in conjunction with chemometrics analysis using PCA and PLS-DA can efficiently detect adulteration in *Ziziphus jujuba* Mill. fruit powder.

2 Materials and Methods

2.1 Sample preparation

The jujube fruits originated from Xinjiang, China were bought from a local market. The fruits were washed to remove any dirt or dust. The fruits were dried, cut into half and the pit was removed. The fruit pulps were evenly distributed on a metal plate covered with aluminium foil with holes and stored in an ultra-low temperature freezer at -80 °C overnight. The frozen fruit pulps were freeze-dried at -105 °C. The fruit pulps were blended, and the powder was freeze-dried. The final product was ground and sieved. The pure jujube powder was stored in a zip lock bag at -80 °C until further analysis.

Adulterated jujube powder was prepared by blending it with corn starch and powdered sugar in varying percentages. First, granulated sugar of 95 g was mixed with 5 g of corn starch and blended to obtain powdered sugar. Adulterated samples containing 2%, 6%, 10%, 20%, 30% and 100% (w/w) of corn starch (CS) as well as 6%, 15%, 30%, and 100% (w/w) of powdered sugar (PS) were prepared. A total mass of 5g for each adulterated sample were mixed in 15 mL centrifuge tubes and vortexed to homogenize. A total of 12 series of samples were prepared, each with three replicates (training data, n = 36). In addition to sample for model building, triplicates of five retail samples (Brand A, B, C, D, E) were collected (testing data, n = 15) from online retailers.

2.2 Fourier Transform Infrared Spectroscopy Measurement

The adulterated samples and the retail samples were scanned using Perkin Elmer Spectrum 100 FT-IR. A total of 8 scans with 4 cm^{-1} resolution from 650 to 4000 cm^{-1} were acquired for each spectrum. Spectrum pre-processing was done to eliminate the instability in IR spectra due to measurement errors and inconsistent compression [8]. Derivative baseline correction, Savitzky-Golay smoothing, and Standard Normal Variate (SNV) normalization was carried out to eliminate scattering using the Perkin Elmer Spectrum Software Version 10.6.1.

2.3 Chemometric Analysis and Model Validation

SIMCA-P version 14.1 (Umetrics, Umeå, Sweden) was employed for chemometric modelling. FT-IR spectra in Microsoft Excel (.xls) format was imported into the software for analysis. All spectral data were mean centred, UV scaled, and auto-fitted before principal component analysis (PCA) and partial least squares discrimination analysis (PLS-DA) were performed. The performance of PLS-DA models was observed through R^2X, R^2Y, Q^2 values and permutation tests. Multivariate calibration analysis was performed using PLS model. The model was evaluated using coefficient determination (R^2), root mean square error of estimation (RMSEE), and root mean square error of cross validation (RMSECV) values [9].

3 Results and Discussion

To get an overview of the data set, PCA was applied to extract meaningful principal components (PCs). The PCA model demonstrated R^2 of 0.972 and Q^2 of 0.907 (Figure 1), indicating a good ability to differentiate the corn starch (lower left quadrant) and powdered sugar (upper left quadrant) adulterants from the pure jujube powder (upper right quadrant). The retail samples, Brands C and D clustered with pure jujube, while Brands A, B, and E clustered with corn starch. This indicated that Brands A, B, and E jujube powders were possibly adulterated with corn starch while Brand C and D jujube powders were not adulterated or contains lower levels of adulterant. The high sugar content of jujube fruit and its tendency to absorb moisture often led to caking or agglomeration [10], which may impact consumer acceptance due to its visual appearance. To improve the appearance of jujube powder, anti-caking agents such as starch, talc, and flour are sometimes added, along with artificial colours or dyes to enhance consumer appeal. Starch is a commonly used adulterant in food products due to its affordability, minimal health impact, and effectiveness as an anti-caking agent [10].

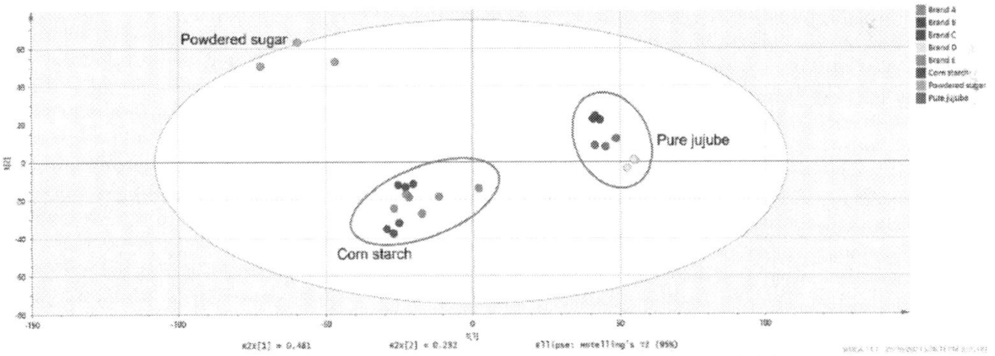

Fig. 1. PCA score plot of different brand of retail jujube powders with pure jujube powder, corn starch and powdered sugar.

Another PCA model was conducted to include only the corn starch series with different levels of corn starch adulteration. The PCA model showed R^2 of 0.981 and Q^2 of 0.936 (Figure 2a). No outliers were observed, suggesting that all the samples are representative of the population. The PCA score plot demonstrated that Brands A, B, and E were separated from Brands C and D by PC1. It was noted that Brands A, B and E form a cluster with the adulterant, corn starch on the right quadrants. This indicated that these three brands shared very similar chemical properties or bonding that transmitted the FTIR light similar to that of corn starch, potentially due to the higher concentration of corn starch in the three jujube powder samples. However, Brands C and D were closely associated with pure jujube powder in the left quadrants, suggesting that these two brands have very similar properties with the pure jujube. Besides, the PCA showed that Brands C and D were separated by PC2, in which Brand C clustered with pure jujube, as well as 2% and 10% corn starch, while Brand D clustered with 6% and 20% corn starch. This suggested that Brands C and D jujube powders may be adulterated with moderate level of corn starch, but the level of adulteration is uncertain. PLS-DA was further used for the discrimination of pure and adulterated jujube powders.

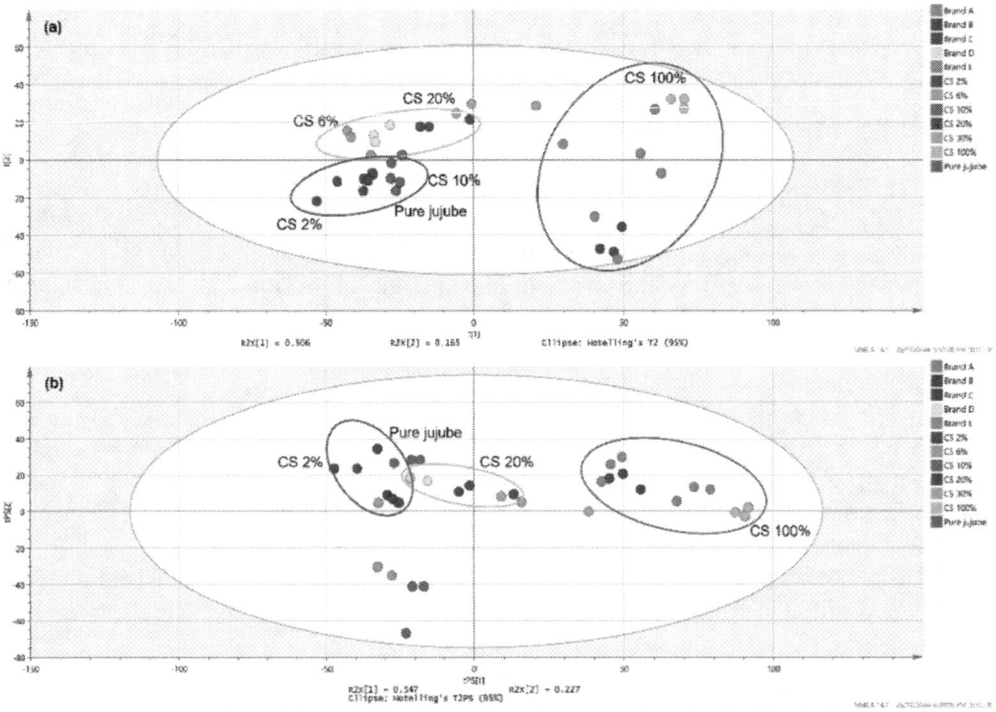

Fig. 2. (a) PCA score plot and (b) PLS-DA score plot of corn starch series for discrimination of pure and adulterated jujube powders. CS, corn starch.

A classification model was built using PLS-DA in order to evaluate the possibility of predicting the level of corn starch adulteration. PLS-DA is a supervised method that uses class labels to maximize the separation between classes [11], in which each Y variable corresponds to a different level of adulteration ranging from 0% to 100%. PLS-DA score plot (Figure 2b) shows that 93% of the variability in the data was accounted for using the first four principal components, indicating a good level of explanation for further analysis. The PLS-DA model also presenting a goodness-of-fit value of $R^2X = 0.977$, $R^2Y = 0.92$, and a goodness-of-prediction value of $Q^2 = 0.738$. A Q^2 value that is > 0.5 and < 0.9 is generally considered good and suggests that the model is not overfitted. Validation using the permutation test resulted in a Q^2 intercept value of zero and below zero indicating model validity. Figure 2b shows that Brand A jujube powder clustered closer to the 100% corn starch, while Brands B and E clustered nearer to 30% corn starch, indicating a higher level of corn starch adulteration in Brand A. Additionally, brand C and D were separated from Brands A, B and E by PC1, as observed in the PCA model. However, the PLS-DA model provides improved discrimination as compared to the PCA model; specifically, PC2 distinguished Brand D as clustering with the 20% corn starch, while Brand C clustered with 0% and 2% corn starch. The results demonstrated that the level of corn starch adulterant in the retail samples follows this order: Brand C (0-2%) < Brand D (~20%) < Brands A, B, and E (30-100%).

Figure 3a shows the PCA score plot of the powdered sugar adulteration series. The PCA model had R^2 of 0.961 and Q^2 of 0.913. The presence of pure powdered sugar adulterant as an outlier suggested that the retail jujube powders may not contain high level of sugar. Brands A, B and E clustered together and were positioned far from Brands C and D, which were clustered with the powdered sugar adulteration series in the right quadrants. Brand C clustered with 15% and 30% powdered sugar in the lower right quadrant and Brand D clustered with 0% and 6% powdered sugar in the upper right quadrant. The PCA model suggested that Brands C and D jujube powders may contain a moderate level of powdered sugar adulteration, which is 15-30% in Brand C and 0-6% in Brand D. However, the clustering pattern could also be attributed to compositional variations among the unknown jujube cultivars used in producing the powder. Additionally, the similarity between the sucrose found in commercial table sugar and the natural sucrose in jujubes [12] may have contributed to the observed clustering pattern. The PLS-DA model of the powdered sugar series (Figure 3b) explained only 88.5% of the variability in the X matrix, and the clustering pattern does not show any improved discrimination from that of the PCA model. The model was demonstrated to be weak and underfitted, owing to the low R^2Y value of 0.391. The low Q^2 value of 0.272 indicated that the PLS-DA model generated using the powdered sugar adulteration series has poor predictive performance and is likely not reliable.

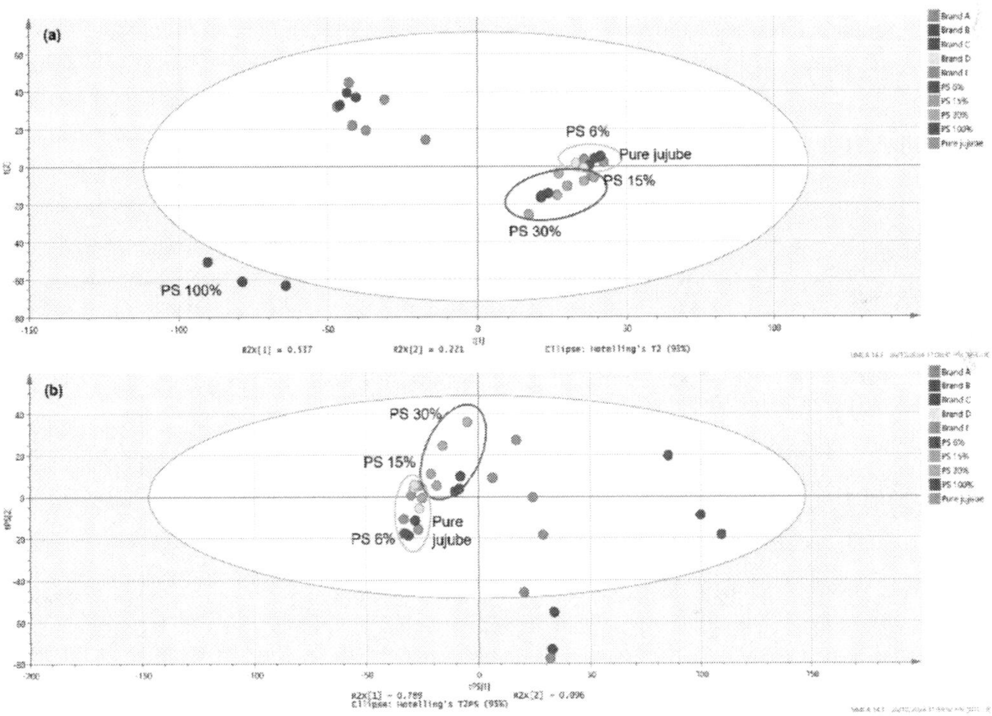

Fig. 3. (a) PCA score plot and (b) PLS-DA score plot of powdered sugar series for discrimination of pure and adulterated jujube powders. PS, powdered sugar.

A calibration model for the relationship between observed and predicted values, as determined using multivariate calibration of the corn starch PLS model was performed. Results showed that PLS was successfully used to detect and predict the concentration of corn starch in jujube powder with high accuracy and good precision. The unknown sample of corn starch adulterated jujube powder could be predicted using the equation of y = x -

5.391×10^{-8} with the R^2 value of 0.9859. In addition, the Root Mean Square Error of Estimation (RMSEE) is low (5.50%), indicated that the model has a good fit to the training data. The calibration model was further evaluated using cross validation, and a low RMSECV (9.16%) was obtained. It can be summarized that multivariate calibration of PLS could be used as rapid and effective methods for authentication of jujube powder by detecting and predicting the concentration of corn starch with high accuracy and precision.

4 Conclusion

This study demonstrated the effectiveness of multivariate analysis, particularly PCA and PLS-DA, in detecting adulteration in retail jujube powders and distinguishing between authentic and adulterated samples. The PCA model successfully revealed clustering patterns that separated adulterated samples from pure jujube powders based on variations in their FTIR spectra, while PLS-DA offered enhanced discrimination and predictability of adulterant levels. Corn starch and powdered sugar adulterations were detected at varying concentrations across different brands, with PCA indicating moderate levels of adulteration for powdered sugar in Brands C and D and PLS-DA accurately estimating corn starch concentration ranges in all five samples. The calibration model developed showed high accuracy and precision, validating the model's robustness in predicting corn starch levels. These findings suggest that PLS-based multivariate calibration can serve as a rapid and reliable method for authenticating jujube powder quality and detecting adulterants, contributing to improved quality control in functional food products.

References

1. N.N. Ahmad, N.S. Khairatun, The understanding of food fraud among consumers in Klang Valley, Malaysia. JoSTE **2**, 31–40 (2021).
2. M.J. Walker, S. Cowen, K. Gray, P. Hancock, D.T. Burns, Honey authenticity: the opacity of analytical reports - part 1 defining the problem. NPJ Sci. Food **6**, 11 (2022).
3. P. Dais, E. Hatzakis, Quality assessment and authentication of virgin olive oil by NMR spectroscopy: a critical review. Anal. Chim. Acta **765**, 1–27 (2013).
4. C. Di Anibal, M.S. Rodriguez, L. Albertengo, UV-visible spectroscopy and multivariate classification as a screening tool to identify adulteration of culinary spices with Sudan I and blends of Sudan I + IV dyes. Food Anal. Methods **7**, 1090–1096 (2013).
5. S. Manasha, M. Janani, Food adulteration and its problems (intentional, accidental and natural food adulteration). Int. J. Res. Finance Manage. **6**, 131-140 (2016).
6. Y. Lu, T. Bao, J. Mo, J. Ni, W. Chen, Research advances in bioactive components and health benefits of jujube (*Ziziphus jujuba* Mill.) fruit. J. Zhejiang Univ. Sci. B **22**: 431-449 (2021).
7. B. Li, M. Yu, W. Xu, L. Chen, J. Han, Comparison of PCR techniques in adulteration identification of dairy products. Agriculture **13**, 1450 (2023).
8. Y. Jiao, Z. Li, X. Chen, S. Fei, Preprocessing methods for near-infrared spectrum calibration. J. Chemom. **34**, e3306 (2020).
9. K.E. Loh, Y.S. Chin, I.S. Ismail, H.Y.Tan, Rapid characterisation of xanthine oxidase inhibitors from the flowers of *Chrysanthemum morifolium* Ramat. using metabolomics approach. Phytochem Anal. 33:12-22 (2022).

10. X. Xia, F. Li, H. Ran, J. Zhao, X. Lei, L. Lei, J. Wen, G. Xiao, K. Zeng, J. Ming, Effect of jujube kernel powder addition on moisture absorption performance, color stability, texture properties and agglomeration characteristics of jujube powder. LWT, **174**, 114452 (2023).
11. M. Lasalvia, V. Capozzi, G.A. Perna, Comparison of PCA-LDA and PLS-DA techniques for classification of vibrational spectra. Appl. Sci. **12**, 5345 (2022).
12. Q. Zhang, L. Wang, Z. Wang, Z. Liu, Z. Zhao, G. Zhou, M. Liu, P. Liu, P. Variations of the nutritional composition of jujube fruit (*Ziziphus jujuba* Mill.) during maturation stages. Int. J.Food Prop. **23**, 1066–1081 (2020).

Mental health, nutritional status, physical activity and healthy eating index among Malaysian university students

HC Koo*, *Jenniffer SY* How, *JS* Koh, *GS* Law and *Elizabeth* Kavisha M

Faculty of Applied Sciences, Tunku Abdul Rahman University of Management and Technology, 53300, Kuala Lumpur Malaysia.

Abstract. Lifestyle is a complex phenomenon that encompasses all aspects of daily life. University is a crucial period when students are more susceptible to engaging in unhealthy behaviours that can harm their health. This study aimed to assess the mental health, nutritional status, physical activity, and Health Eating Index (HEI) among university students at a university in Klang Valley, Malaysia. In this cross-sectional study, body composition (weight, height, body fat percentage and visceral fat), physical activity (IPAQ questionnaire), healthy eating index (HEI-Malaysia questionnaire) and mental health (GHQ-12 questionnaire) were collected from 383 university students aged 19-30 years using purposive sampling method. Approximately half proportionate of them (n=172; 45%) had unhealthy HEI scores, and 15% of them were overweight/ obese (n=57). However, a majority of them had high physical activity level (n=276; 72.5%). Physical activity was inversely associated with body fat percentage (r=-0.101; *p*-value<0.05). Meanwhile, GHQ-12 scores were significantly associated with HEI scores (r=0.211; *p*-value<0.01) and body mass index (r=-0.107; *p*-value=0.036). The findings of the study indicate that university students are prone to unhealthy eating patterns. Those with good mental health tend to practice healthy eating and have a lower body mass index.

1. Introduction

1.1 Background

The transition to higher education presents numerous challenges for students, significantly impacting their overall well-being and academic success [1]. University students in full-time higher education typically range in age from 17 to 24 years, which is also a critical period for the onset of psychological issues. The transition from home to university can impact mood and overall mental health. This is a period of increased risk of onset of mental health problems. Studies from various countries have indicated that the well-being and mental health of students are under considerable pressure, this presents a serious health and societal issue [2]. Hence, the well-being of students in higher education is receiving increasing attention.

* Corresponding author: koohc@tarc.edu.my

The transition from home to university also appears to impact students' diet quality, as this period is often marked by the adoption of poor dietary habits [3]. Research indicates that poor dietary quality contributes to a greater risk of metabolic diseases. The issues of obesity and overweight among university students are alarming. Meta-analysis has revealed that university students tend to gain an average of 1.6 to 3.0kg during their four-year enrolment [4]. This increase in weight is five times higher than the average weight gain projected for the general population in one year. Obesity subsequently contributes to a higher mortality rate from a range of chronic diseases. Physical activity also plays a crucial role in determining the health status of a young adult, besides dietary intake. Given that a majority of the world's population is physically inactive, this issue is regarded as a public health concern rather than just an individual issue. Among the most significant external factors impacting physical activity in the student population is the lack of free time, attributed to faculty schedules and social and family responsivities, adversely influencing students' well-being, strength, endurance and nutritional status [5]. It is widely recognized that physical activity has numerous positive effects on overall health, including both physical and mental well-being [6]. Thus, maintaining an active lifestyle is essential.

A growing number of studies have recently examined the potential impact of diet quality on the development of mental health issues across various nations and cultures [7]. Though not the only contributing factor, poor diet quality is recognized as a risk factor for mental health problems, with some arguing that mental health issues can also influence diet quality in young adult population [8]. Besides, high levels of physical activity are associated with mental well-being and social well-being, in addition to physical well-being [5]. As university students navigate the transitional phase in their lives, they often encounter changes in their living habits, moving from the structured setting of high school to a more flexible university environment. It involves challenging responsibilities, unfamiliar situations and more complex academic tasks; this suggests that university students are at risk and experience a high prevalence of mental health issues [9]. However, the associations between mental health, nutritional status, healthy eating index and physical activity level in Malaysian university students are not being studies. Considering all the above, it is important to investigate the mental health, nutritional status, healthy eating index and physical activity level among the university students in Malaysia.

2 Methodology

2.1 Participants recruitment and study design

A total of 383 healthy university students aged 19-30 years from a private university in Kuala Lumpur were recruited for this cross-sectional study through purposive sampling. This study was granted ethical approval by the Research Ethnics Committee of Tunku Abdul Rahman University of Management and Technology, and was conducted in line with the guidelines laid down in the Declaration of Helsinki. Prior to data collection, written informed consent was obtained from all the participants. The sample size was calculated using the krejcie & Morgan [10] formulation. Given that the university has a total of 20,000 university students, with a relative precision of 5%, and predicted prevalence of 50% for a confidence interval of 95%. The estimated sample size needed for this study was 377 participants. We included an extra 10% to the sample size to account for the expected non-response rate, and the possibility of information loss resulting from incomplete data, the final required sample size for the present study was 415. The inclusion criteria were: (1) apparently healthy Malaysian university students aged 18 years and above; and (2) must be able to write, read and understand English language. Those who were absent on the data collection day or had serious comorbidity requiring treatment were excluded from the study, as their presence could potentially affect the outcomes.

2.2 Anthropometric measurements and body composition

Each participant's body weight was measured twice using a calibrated InBody 270 (InBody Co. Seoul, Korea) and recorded to the nearest 0.1kg. The participants were weighed while clothed, barefoot and without belts or items in their pockets. Participants were required to stand still, ensuring their body weight was evenly distributed on both feet. The angle of their heads had to be in the Frankfurt plane, with an upright posture, upper limbs at their sides and palms facing forward, feet flat and directed forward toward the researcher. Height measurement for each participant was conducted twice using a BSM170 Stadiometer (InBody Co. Seoul, Korea), recorded to the nearest 0.1cm. Participants were instructed to stand erect, barefoot, with their heels, buttocks, head and shoulder blades in alignment against the wall. Measurements were taken in the upright position, ensuring the top of the external auditory meatus was level with the inferior margin of the orbital bone. Body mass index was calculated in accordance with World Health Organization (WHO) guidelines.

2.3 Mental health assessment

In the present study, the 12-item General Health Questionnaire (GHQ-12) was employed to assess the psychological health of university students. This self-administered screening tool aimed to identify psychological morbidity within the general population and those who may be at risk of psychiatric disorders. The GHQ-12 consists of 12 items scored using the original scoring method. 4-point Likert scare was adopted. Positively worded items have responses of "better than usual", "same as usual", "less than usual", and "much less than usual". Whereas, negatively worded items were scored with "not at all", "no more than usual", "rather more than usual" and "much more than usual". The total GHQ-12 scores ranged from 0 to 36, with higher scores indicating an increased risk of psychiatric comorbidity. In the Malaysian version of the GHQ-12, the optimal cut-off points for identifying individuals at risk of poor mental health is set at ≥ 9 [11].

2.4 Healthy eating index

The Malaysian Healthy Eating Index (HEI) serves as one of many metrics for assessing the adherence of Malaysian citizens to the dietary recommendations set forth by the Malaysian Dietary Guidelines (MDG) [12]. The questionnaire comprises nine components: seven dietary categories and two nutritional groupings. These components yield various scores that contribute to an individual's overall dietary composition. Each nutrient's score ranged from 0 to 10, with whole number responses tallied proportionally. The overall healthy eating index score was calculated by summing the scores of each component. The composite score as a percentage was determined using the formula: (total score from food components/ maximum score) x 100%. A total score of 100% is possible, where a score of less than 46% indicates a poor diet and more than 46% indicated a good diet [12].

2.5 Physical activity assessment

Physical activity level was assessed through the self-administered International Physical Activity Questionnaire (IPAQ). Participants were asked 27 questions related to their activities during the past week, including leisure, house chores and gardening, work-related activities, and transport-related physical activities. The IPAQ employs a scoring method in which each activity is assigned a Metabolic Equivalent (MET) value to each activity. MET represents the ratio of the metabolic rate during an activity to the metabolic rate while at rest. The MET score for each activity is calculated as follows:

Walking MET score = 3.3 x walking minutes x days of walking
Moderate MET score = 4.0 x moderate activity minutes x days of moderate activates
Vigorous MET score = 8.0 x intense vigorous activity minutes x days of vigorous activities

The total MET score for physical activity was determined by adding the walking, moderate and vigorous MET scores. Subsequently, the scores were classified into low, moderate and vigorous physical activity levels according to below category:

Low physical activity level : $<$ 600 MET minutes/ week
Moderate physical activity level : 600-2999 MET minutes/ week
High physical activity level : $\geq$ 3000 MET minutes/ week

2.6 Statistical analyses

SPSS (version 22.0) (IBM SPSS Statistics, 2014) was used for data analysis. Before analysis, the data was entered, cleaned and thoroughly checked. Descriptive statistics were computed for participants' demographic information, anthropometric profiles, GHQ scores, HEI category and physical activity level. The Kolmogorov-Smirnov test was used to assess the normal distribution of each variable. Group differences were tested using the Chi-Square test and the Mann-Whitney U test. The associations between mental health, nutritional status, HEI scores and physical activity were determined using Spearman's correlation coefficients. A two-sided p-value of <0.05 was considered as statistically significant.

3 Result

The demographic information, anthropometric profiles, GHQ scores, HEI scores and physical activity MET scores among participants are presented in Table 1. A total of 383 university students have been enrolled into the present study. A majority of them (65.3%; n=250) were normal weight, with normal body fat percentage (54.8%; n=210) and normal visceral fat (93.7%; n=359). 15% of them were overweight/ obese (n=57). Approximately half proportionate of them (n=172; 45%) had unhealthy HEI scores. However, a majority of them had high physical activity level (n=276; 72.5%). Male showed significantly higher physical activity MET score as compared to female (p-value<0.01).

Table 1. Demographic information, anthropometric profiles, GHQ scores, HEI category and physical activity level (n=383)

Variables	Male (n=174)	Female (n=209)
Age (years), median (IqR)	21 (1)	21 (2)
BMI (kg/m^2), median (IqR)	21.3 (4.4)	20.1 (3.8)
Body fat percentage (%), mean (SD)	19.2 (8.0)	30.3 (7.3)

Visceral fat, median (IqR)	4.0 (4.0)	5.0 (4.0)
Healthy eating index score, median (IqR)	47.4 (19.2)	48.8 (16.7)
GHQ-12 score, median (IqR)	13.0 (6.0)	14.0 (8.0)
Physical activity MET score, median (IqR)	5951.8 (13745.3)	3879 (9483) ††

IqR: Interquartile range; Mann-whitney U test was applied, ††*p*-value <0.01

Table 2 shows the associations between mental health, nutritional status, healthy eating index and physical activity among the university students. Physical activity was inversely associated with body mass index (r=-0.130; *p*-value<0.05) and body fat percentage (r=-0.101; *p*-value<0.05). Meanwhile, GHQ-12 scores were significantly associated with HEI scores (r=0.211; *p*-value<0.01) and body mass index (r=-0.107; *p*-value=0.036).

Table 2. Associations between mental health, nutritional status, healthy eating index and physical activity among the university students (n=383)

	BMI	**% BF**	**VF**	**HEI**	**MET**	**GHQ**
BMI		0.624††	0.075††	0.130	-0.130†	-0.107†
% BF	0.380††		0.895††	0.042	-0.101†	0.013
VF	0.624††	0.895††		0.085	-0.020	-0.027
HEI	0.075	0.042	0.085		0.080	0.211††
MET	-0.130†	-0.101†	-0.020	0.080		-0.040
GHQ	-0.107†	0.013	-0.027	0.211††	-0.040	

BMI: body mass index; % BF: percentage of body fat; VF: visceral fat; HEI: healthy eating index; MET: physical activity MET score; GHQ: general health questionnaire; ††*p*-value <0.01; †*p*-value <0.05.

4 Discussion

This is the first study to investigate the associations between mental health, nutritional status, healthy eating index and physical activity among Malaysian university students. Our study reveals the significant inverse associations between physical activity with body mass index and body fat percentage, as well as significant positive associations between healthy eating index and body mass index. In the present study, male demonstrated a significant higher physical activity MET score as compared to female. This gender disparity, favouring men, has been corroborated by previous study [13]. Recent research indicates that men demonstrate higher levels of motivation towards physical activity across various field, with the exception motivation and demotivation, possibly driven by competitive aspects and immediate rewards. This aligns with research indicating that competition can enhance motivation, particularly for men. On the other hand, women's higher levels of external motivation and demotivation suggest that social factors or expectations might play a significant role in their engagement with physical activity. The preference for long-term goals could indeed contribute to a more sedentary lifestyle, especially if those goals are perceived as challenging to achieve in the short term [14].

Obesity is both a direct and indirect cause of serious chronic diseases. Many studies over the past few decades have sought solutions to the problem of obesity. Physical activity has been found to be inversely associated with body composition, it is in accordance with most longitudinal studies, and this observation has been interpreted as evidence supporting the protective role of physical activity in mitigating excess body fat accumulation among youth [15]. Besides, our study also found the inverse relationship between body mass index and mental health, it is in line with systematic review and meta-analysis, which demonstrated a bi-directional relationship between obesity and mental health [16]. While several studies examine the relationship between obesity and mental health, none have specifically investigated this issue among university students.

Recently, research has increasingly focused on the potential association between diet quality on the emergence of mental health issues across different cultures and countries. Many studies have found that eating a healthy diet can lower the risk of mental health issues [17]. Similarly, our study also demonstrates the same outcome among the university students. One study has associated the consumption of fruits with mental health, this relationship could be attributed to the various benefit nutrients found in fruits, including vitamins, folate, antioxidants, phytochemicals and minerals, which may play a protective role against psychiatric disorders. Additionally, the carbohydrate in fruit can influence insulin secretion. This process may enhance the production of neurotransmitters like serotonin by facilitating the transport of tryptophan to the brain. Increased serotonin levels are often associated with improved mood and emotional well-being. Overall, incorporating more fruits into the diet could be a simple yet effective strategy to support mental health [17].

On the whole, findings of the present study indicate that those with good mental health tend to practice healthy eating and have a lower body mass index. The university setting is ideal for implementing health promotion programs. Therefore, it is crucial to execute relevant initiatives that motivate students to take responsibility for their well-being and adopt healthy eating habits, along with other forms of wellness.

References

1. Rodriguez, Neisha M and Gil, Karen. Eating Habits, Health Status, and Emotional Well-being Among an Optometry Student Population. Health Professions Education. **9**, 3 (2023).
2. Backhaus, I., Varela, A. R., Khoo, S., Siefken, K., Crozier, A., Begotaraj, E., Fischer, F., Wiehn, J., Lanning, B., Lin, P., Jan, S., Zaranza Monteiro, L., Al-Shamli, A., La Torre, G., & Kawachi, I. Associations between social capital and depressive symptoms among college students in 12 countries: Results of a cross-national study. Frontiers in Psychology. **11** (2020).
3. Ramón-Arbués E., Granada-López J.M., Martínez-Abadía B., Echániz-Serrano E., Antón-Solanas I., Jerue B.A. Factors Related to Diet Quality: A Cross-Sectional Study of 1055 University Students. Nutrients. **13** (2021).
4. Bailey CP, Sharma S, Economos CD, Hennessy E, Simon C, Hatfield DP. College campuses' influence on student weight and related behaviours: A review of observational and intervention research. Obes Sci Pract. **6** (2020)
5. Podstawski R, Finn KJ, Borysławski K, Omelan AA, Podstawska AM, Skrzypczak AR, et al. The Influence of COVID-19 on University Students' Well-Being, Physical Activity, Body Composition, and Strength Endurance. International journal of environmental research and public health. **19**, 12 (2022).

6. Mahindru A, Patil P, Agrawal V. Role of Physical Activity on Mental Health and Well-Being: A Review. Cureus. **15**, 1 (2023).
7. Açik M, Altan M, Çakiroğlu FP. A cross-sectionally analysis of two dietary quality indices and the mental health profile in female adults. Current Psychology. 2020
8. Lyzwinski LN, Caffery L, Bambling M et al. The relationship between stress and maladaptive weight-related behaviors in college students: a review of the literature. Am J Health Educ. **49** (2018).
9. Harrer M., Adam S.H., Baumeister H., Cuijpers P., Karyotaki E., Auerbach R.P., Kessler R.C., Bruffaerts R., Berking M., Ebert D.D. International Journal of Methods in Psychiatric Research. **28** (2019).
10. Krejcie, RV & Morgan, DW. Determining sample size for research activities. Educational and psychological measurement. **30** (1970).
11. Yusoff, MSB. The validity of two Malay versions of the General Health Questionnaire (GHQ) in detecting distressed medical students. ASEAN J Psychiatr. **11** (2010).
12. Goh H.W., Norimah A.K. Validation of Healthy Eating Index (HEI) for Malaysian adults; Proceedings of the 27th Scientific Conference of the Nutrition Society of Malaysia; Kuala Lumpur, Malaysia. (2012).
13. Fan, X.; Zhu, Z.; Zhuang, J.; Liu, Y.; Tang, Y.; Chen, P.; Cao, Z. Gender and age differences in the association between living arrangement and physical activity levels among youth aged 9-19 years in Shangai, China: A cross-sectional questionnaire study. BMC Public Health. **19** (2019).
14. Ramires VV, Dumith SC, Goncalves H. Longitudinal association between physical activity and body fat during adolescence: a systematic review. J Phys Act Health. **12**, 9 (2015).
15. Jokela M, Laakasuo M. Obesity as a causal risk factor for depression: systematic review and meta-analysis of mendelian randomization studies and implications for population mental health. J Psychiatr Res. *163* (2023).
16. Qiao Y, Wang C, Tian X, Cao M. Association between loneliness and mental health among nurses: a cross-sectional study in China. BMC Health Serv Res. **24** (2024).
17. Galbska D, Guzek D, Groele B, Gutkowska K. Fruit and vegetable intake and mental health in adults: a systematic review. Nutrients. **12** (2020).

AUTHOR INDEX

EDP Sciences – Web of Conferences
17, Avenue du Hoggar
Parc d'Activité de Courtabouf
BP 112
F-91944 Les Ulis Cedex A
France

ISBN 979-8-3313-1550-4
9 798331 315504

CURRAN ASSOCIATES INC.
proceedings.com

11th Southeast Asia Astronomy Networking Meeting (SEAAN Meeting 2019)

EPJ Web of Conferences Volume 240 (2020)

Singapore
26 - 28 December 2019

edp sciences